Natural Cattle Care

Pat Coleby

Acres U.S.A.
Austin, Texas

Natural Cattle Care

Copyright © 2001, 2010 by Pat Coleby

Acres U.S.A.
P.O. Box 91299
Austin, Texas 78709 U.S.A.
(512) 892-4400, fax (512) 892-4448
info@acresusa.com • www.acresusa.com

Printed in the United States of America

Publisher's Cataloging-in-Publication

Coleby, Pat
    Natural cattle care/Pat Coleby
    xii, 220 p., 23 cm.
    Includes index.
    ISBN: 978-0-911311-68-6

    1. Cattle. 2. Cattle — Nutrition — Requirements.
    3. Cattle — Health. 4. Cattle — Breeding. I. Title

SF 197.C65 2000   636.2'08

Contents

Preface

This launches the first United States edition of *Natural Cattle Care*. I am grateful to Fred Walters and his family for giving me this opportunity to present a new version of my book on cattle to a new country.

"A clever person solves problems, a wise one prevents them."

This quotation was used by Dr. Stuart Hill of McGill University, Quebec, and now Sydney University, to open a talk on organic agriculture at a seminar held at Camden, New South Wales. I think it describes very well my approach to agriculture in general and animal care in particular.

Prevention is better than cure is a recurring theme in this book and I hope it will help cattle farmers to think along those lines rather than assuming the many ills that can beset their cattle are a normal process.

My obsessive interest in the processes that make animals sick and ways of preventing this happening by improving husbandry from the ground up, has led me a long and interesting way. Dr. William Albrecht said, "Feed the soil and not the plant." When the soil is healthy, the plant follows suit, and every person and animal that lives off the plants and their products is as well.

The stark facts of today's agriculture are that we are trying to stop a downward spiral of soil and land degeneration due to extended use of chemical fertilizers and sprays. The lack of appreciation of the soil deficiencies inherent in Australia has made this process worse. Unfortunately, the problems of soil deficiencies are no longer confined to Australia — other countries have reduced extremely good soil to something near a disaster in many cases.

Farmers are coming to realize the hard way that all is not well. The road to this state of affairs was long; the one back from it is comparatively short *if* we all take the necessary steps. Hopefully this is starting to happen and the improved health of our cattle and all farm stock, as well as our own good health, will ultimately be the reward.

Pat Coleby, 2000
Maldon, Victoria, Australia

Chapter 1

Farming in the New Millennium

In the new world, the United States and Australia, new methods of farming have been employed during the last century and not all of them have been an improvement. Now that we are into the 21st century, so much new technology is available that one wonders whether the cattle are still the same. Some have remained the same, but many more are not.

Cattle in Early United States History

In a book written in 1955 about the American Indians and white settlement in the United States, the opening up of what may have seemed like an endless cattle kingdom is just another story of how we somehow went wrong. Helped by the best natural pasture possible, home of the bison until they were killed out, the feudal cattle kingdoms inherited from the Spanish took over the land. Populations of longhorn cattle increased until, at last, they reached a point where the numbers far outreached their original purpose as food for the settlers of the region.

These huge spreads eventually were divided into large ranches and a new factor, the western cattleman, was well established by the time of the Civil War. As human populations increased, the skills of the cattlemen enabled huge numbers of cattle to be moved along the new railways to provide an endless supply to meet the growing demands of the new towns. The cattlemen and their charges often faced incredible hardships in achieving this object, and thus arose a whole culture — widely, and generally erroneously, copied by the movie moguls in due course.

The American prairies and cattle country were apparently stronger than similar types of country in Australia and lasted for many years; but inevitably the grass ceased to grow so well all the year despite the fact that the buffalo herds had been much reduced. Between about 1883 and 1886, nature struck back. Blizzards, droughts and incredible winds cost the lives of thousands of cattle. The sheep ranchers had also started to move in bringing the twenty-year bonanza in cattle to an end. The plains were never quite the same again and fences started to figure more and more.

Homesteaders and old-fashioned farming

At the beginning of the 20th century, farming generally followed an established pattern in the older, settled countries of the world. This system was similar in most respects to that which had been in use during the previous century. Some strides had been made in the 1850s when Hensel discovered that rock dust, in his case something very similar to what we know as dolomite today, was needed on ancient and over-manured pastures to help them regain their fertility.

The British farmer who taught me in the middle of the last century explained that, when lime levels are not attended to, land can become just as sick from over-manuring with a natural product as it can from gross deficiencies. It becomes sicker even more quickly when excessive amounts of N, P, and K fertilizers are used, and so do the cattle that run on the land.

Nowadays, rock dust is being used in many countries to remineralize soils. Australia has rather poor quality rocks which means that our soils are generally lower in quality when compared with the rest of the world. This is quite unlike Canada and the United States which have high-quality rocks, like those that are being used for remineralization by Hamaker in Canada, for example. Basalt (blue stone) rock dust is possibly the best quality we have here, and it does contain some calcium and magnesium. Checking out your local quarry and getting a readout of the minerals in the rocks is always a good idea. Sometimes the dust can be used as a low-cost topdressing at the rate of one-half to one ton to the acre.

Fencing

Modern fencing is perhaps the single most important item in revolutionizing farming practices and enabling cattle to be managed more easily and efficiently. "Folding" animals on crops, so that their grazing could be controlled, is an old farming practice. Up until the advent of wire fences, it was done by labor-intensive methods involving wooden hurdles and manpower — not easy on a large scale. Otherwise, in many countries human herders were used to control grazing cattle. Strip grazing systems with electric fences are now an integral part of any large scale cattle farming and pathways for moving cattle, instead of pushing them from one paddock to another, have made life much easier. Electric fencing systems on their own are used on some farms, but even nowadays cattle are notorious for going where they want and electric fences, except in controlled strip grazing, do not always act as deterrents. Well-made, plain wire fences, if properly erected, control cattle as well as possible.

We are also seeing quite a few farmers returning to André Voisin's methods, mostly called cell grazing or management-intensive grazing these days. This is a complex system where paddocks are divided into smaller cells and grazed intensively for shorter periods of time in order to promote better grass growth, higher soil quality, faster manure breakdown, and improved animal health among other things. This system is made much easier with modern fencing materials.

The advent of concrete on farms

Concrete is now mandatory for easy and relatively hygienic cattle management. At the beginning of the 20th century, the cost of concrete and the difficulty of making it in large quantities made its use a rarity. Now every dairy farm is made as easy to manage as possible with the aid of concrete. Concrete needs to be used with discretion however.

There have been troubles in the United States with the huge new concrete complexes used for housing cattle in the winter. These were made to replace the old wooden barns which have been used ever since settlement and worked remarkably well. Electricity, especially when used for in-building heating, has produced some strange results when it has not been properly grounded and apparently quite a few farms have had to rethink some of their plans.

One enormous shed in New England, built to replace wooden barns, had to be vacated in nine months because the cows became so ill that they ceased producing. That mistake proved very costly. These are teething troubles that have to be addressed. When a new material becomes available, farmers, like other people, use it to the exclusion of all else, a marriage of old and new can often work better.

Modern milk and meat requirements

Perhaps the biggest changes have been in the type of product now considered desirable. For years fat was regarded as being one of the main objects of beef cattle breeders. The very shape of the animals tended to be quite different from the better muscled and more rangy bovine we see in Australia today. Herds moved by early Americans would probably have been fairly similar. In the dairy industry as well, fat was regarded as the most desirable factor and solids-not-fat and proteins were hardly considered.

The lifestyles of human populations have changed and with it their health. This is due in part to the fact that physical work is not so arduous as it used to be. Hard work helps to work off fat (perhaps not everyone would agree with this assessment on my part). Fat is no longer an important means of keeping warm, so the fat content of meat and milk is no longer the primary concern. It is also now realized that humans do not require so much fat and in excess, without sufficient exercise, it contributes to ill health. As with all fashions we seem to have swung too far in our fat-free obsessions, and a little is actually healthful. Butter is still needed, even if much of our cooking is done using vegetable oil instead of beef drippings. Modern milk requirements are for yogurts, quarks and cheeses.

Good milk protein levels and solids-not-fat for the whole-milk and cheese industries are the most desired qualities these days in dairying. Many soils have become denatured due to the use of artificial fertilizers and this fact makes achieving these goals quite difficult. Farmers will have to look to the state of the land that grows their feeds to remedy the situation. Low protein in the feed is generally caused by soil degeneration. Demineralized soils mean low proteins in the produce. Proteins in wheat alone have fallen from around 15 to 20 percent to as low as six percent in some areas of Australia and possibly also the United States during the last 60 years. In Europe, 18 percent protein is regarded as normal.

Advent of many new and composite breeds

In cattle breeding there are now a host of new, at least new to Australia, composite breeds. These are being added to our basic and durable Angus and Herefords, many of them very successfully. Australian cattle breeders have not been slow in establishing this trend; Droughtmasters, Brafords, Australian Reds, Murray Greys, Illawarra Shorthorns, to name a few, have been added to our breed pool. A great many exotic importations from Europe, the United States, Africa and even Japan are now coming into the country. The latter, generally Waygu and similar types, have been bred for generations in Japan to suit Japanese preferences. They are catching on here and the Japanaese are very pàrticular in insisting on organic production.

The regulations on imports have been reviewed to help streamline procedures to the point where bringing in new genes (either as semen or individuals) for established breeds, or starting up entirely new (to Australia) breeds such as the Belgian Blues, etc., is comparatively easy. Unfortunately, the advent of BSE (bovine spongiform encephalopathy or mad cow disease) in the United Kingdom has closed many markets, and even now the situation is not fully resolved. By sheer luck Australia seems to have escaped, the United States did not.

Comparisons of management systems

The European, North American and Canadian management systems do not have to be followed in Australia or the southern parts of the United States since the climate is warm and it is rarely necessary to yard cattle during the winter. They can go out year round, except in very rare years. In Australia, the spring of 1992 was recorded as the wettest since 1916 and many cattle farmers were caught by surprise as 36 inches of rain fell in two months. They did not realize the damage that "pugging" by a multitude of cattle hooves could do and failed to yard their cattle. Some of those fields are still a mess. This came as a shock to many cattlemen who had not seen farming in action in the northern hemisphere. I advised many of my farmers to sacrifice a smaller paddock and virtually create a feedlot until the ground dried out.

Irrigation and feedlots

We have irrigation systems set up for milking concerns that, when well managed, are extremely efficient and the cattle do well. However, when run badly they are disastrous both economically and ecologically. I feel that a rule of thumb should be that these systems should use water stored on-farm, not obtained at the expense of rivers that cannot supply enough to keep their natural water flow. This has become a problem in Australia. All stock actually prefer grass that is not watered if they have a choice, as I discovered when I had an irrigation farm. The horses and goats preferred dried off feed to the green, watered paddocks.

Intensive systems like feedlots are being used extensively for beef cattle. The steers are bought in as weaners and grown to killing weight in a comparatively short period of time. The initial problems of pollution to waterways and underground aquifers due to manure runoff have been sorted fairly effectively here in Australia. Most concerns now compost their manure for two years or more, after which time tests have shown that undesirable substances, like cadmium and drugs, are no longer present. These concerns are now providing farmers, organic or otherwise, with a cheap and very good source of extra humus — much needed on our farmlands since most of them are lacking in organic matter. Feedlots are a fact of life at present. Some countries, Sweden for instance, are phasing them out, but it is difficult to see that happening worldwide in the foreseeable future.

Unfortunately, managing a large feedlot on organic principles would be very difficult to impossible. Cattle, like all stock, need exercise to stay fully healthy and therefore produce healthy meat. I do know several farmers who run on about 20 or 30 head at a time organically and do it very well. The damage to the environment does not take place in these concerns, and the beasts are only brought in to top them off for the last month or so.

Increased demand for naturally-produced meat and milk

There is now a growing demand for well-grown, uncontaminated produce. This is a worldwide trend for meat, milk and cheeses. People are realizing the benefits of healthy food. Already there is an appreciable number of farmers producing organic beef on a large scale, using the same marketing procedures as conventional farmers. On the dairy scene, there are a farms providing organic milk to the regular market, and the demand organic dairy products is increasing.

Not surprisingly, considering conventional farming has been the normal procedure for almost three generations of farmers, there is a certain amount of apprehension about changing to this type of farming. However, a significant number of farmers are now considering the options quite seriously and find it is not as difficult to put an organic operation in place as people think. When properly managed on regenerated land, it ultimately will be found to be cheaper in many ways. Saving on the cost of expensive drugs and hormones needed to keep conventionally-grown animals producing at high levels, and using natural minerals to keep them in optimal health, is actually quite easy. One big dairy farmer in the United Kingdom, who did a total turn around in six months, wondered how he could have been led astray for so long.

Differences between organic and conventional products

In the United Kingdom during the 1940s and 1950s, tests were done on Lady Eve Balfour's Haughley Experimental Farm in southeastern England. The tests were conducted on crops grown on the farm and the experiment was overseen by a German soil scientist named Schuphan. The farm was divided in two, half was run organically (called mixed farming at that time) and the other half was farmed conventionally.

The same program of cropping, meat and milk production was used on both halves and the produce of each was tested once every month for 15 years. The output from each half was roughly the same, but the difference lay in the quality of the products. Those which came from the conventional half were found to be 28.5% lower in all minerals and vitamins compared to those gown on the naturally farmed half. That is nearly one-third lower in nutritional quality — so it is hardly surprising that human and animal health is not all it should be.

Potassium was the most serious problem found on the conventional section. On the organic half, this was naturally replaced in more than adequate amounts each year, but on the conventional program the potassium got steadily lower, in spite of efforts to boost it artificially. Neither muriate nor sulfate of potash do anything except raise the salt table to undesirable levels. Both are at best a band-aid solution.

From this experiment it was deduced that farming systems that use chemical fertilizers exclusively need to be reviewed because when potassium runs out, farming ceases to be viable. In Australia, we are getting somewhere near that stage, as we already have far more potassium deficiency conditions than other parts of the world, both in animals and humans. Judging by some of the information I receive from the United States, they are not far behind.

Changes needed in the 21st Century

As we begin this new century, we have a chance to turn the tide on chemical farming. I have been on excellently run conventional farms where the chemical fertilizers were used minimally and great attention was paid to the lime levels as well, thus keeping the calcium and magnesium at a good level.

One factor will have to addressed and that is the falling sulfur levels in the soils. This means a host of undesirable diseases in cattle and other stock. In 1998, Neal Kinsey told me that sulfur shortfalls were the fastest growing deficiency in the States at that time. The proliferation of oilseed crops is partly the cause. *Acres U.S.A.* reported some years ago that an oilseed crop could take 50 pounds of sulfur to the acre out of the fields. Sulfur is one of the many minerals inhibited by N-P-K fertilizers.

Regrettably, I have seen organic farms which were not well run and that have fallen into the trap of thinking they did not need to maintain their land in any way. They were organic farming by default as it were, not using artificials, or anything else that was needed to maintain the mineral and organic balance of the soils. However, there are now some very fine, well-managed, naturally run concerns that are producing a high-quality commodity, whether crop, meat or milk. The expertise is there and working well.

Range and mountain cattle husbandry

These types of husbandry have possibly changed less than any other kind of farming. The big cattle spreads in northern Australia are much the same now as when they started in the last century. For many years, they were run on fairly similar lines to the old cattle country in the United States. Mechanization has replaced the horse in some areas, and more attention is paid to mineral requirements these days on many of the stations, but the actual management has changed little.

Mountain cattle farming is still with us. Fortunately, this is a method that is eminently satisfactory from all points of view. The cattle are de-pastured in the mountains during the warmer summer months, rather as they are in Europe, and then brought down to paddocks below the snow line in winter. These paddocks are often kept strictly for this purpose so they have a breather during the summer, leaving them in good heart for winter feed. This system thus has a twofold advantage; it ensures the maintenance of mountain districts where fires will not be an all consuming disaster if they do occur, and provides good, fresh keep for the beasts in the summer.

The cattle have been excluded for a few years from areas like the national park areas in southeastern Australia. The vegetation there is now thick and out of control, often due to European plants like blackberries and gorse which have taken over, that when it burns (given Australian conditions it will be *when* not *if*), the consequences will be an ecological disaster. This scenario is also applicable in other parts of the world where white people's farming methods and intervention have upset the natural balance.

In areas where there is no control of introduced plants, the fire will be so hot and burn so deeply that nothing will grow or regenerate and trees may be permanently lost. This is what is believed to have happened to the Keilor Plains northwest of Melbourne (in southeast Australia) five hundred years ago. A burn-off by the aboriginals of the time got out of control. It is only now with closer settlement and modern tree planting efforts that the trees are being returned to that area. All the natural seed is consumed in conflagrations of that kind.

There are also large numbers of very beautiful wildflowers to be found in our mountain regions where the cattle are now run and it would be terrible if they were lost. The cattle keep them from being choked out by opportunist weeds and plants. In the 1980s, areas of the Wombat Forest north of Bacchus Marsh in southern Victoria burned so hot that no seed was left there to restore the natural bush afterwards.

Farming trees and cattle together

Cattle and trees go well together. In fact cattle, like all stock, will do better if they have shade and shelter when they need it. I am reminded of a friend who bought an area of virgin bush to run beef cattle on — unfortunately we did not meet before he started his clearing program. He clearcut half of it, at which stage he ran out of money, so he had to put the cattle out on the area as it was. To his dismay, the cattle disappeared into the uncleared half and were totally uninterested in the carefully sown and "improved" cleared land. He has been regretting his haste ever since. His animals were looking for copper and other minerals. Cattle, like other stock, only damage trees when deficient in copper and if adequate copper and other minerals are provided in their licks, it is enough to stop them creating mayhem on the land.

There are good books out on agroforestry and trees on farms. One, by Rowan Reid and Geoff Wilson, called *Agroforestry in Australia and New Zealand,* is required reading for any farmer, cattle or otherwise (there are, I am certain, many other books on growing trees with cattle). Trees are a crop like any other and they can be successfully combined with cattle farming of any kind.

Chapter 2

Cattle for Milk and Meat Production

At one of the first Royal Shows in Melbourne after World War II, the only breeds to figure were Beef Shorthorns, Poll Shorthorns, Dairy Shorthorns, Herefords, Polled Herefords, Aberdeen Angus, and Devons (only just arrived apparently, six classes and no entries). There were also a few Red Polls, Australian Illawarra Shorthorns, Ayrshires, Jerseys, Guernseys and mainly British Friesians. The picture has certainly changed.

Unfortunately the advent of BSE (bovine spongiform encephalopathy) in the United Kingdom in 1988 has upset the export of British breeds considerably. It is possible that we shall have to manage with the genes we already have from that part of the world for the foreseeable future.

Milk breeds

Australian Red

This breed, whose origins are discussed in Chapter 5, is becoming established here and is accepted as a part of the dairy scene. There is much to recommend them, not the least is a large and diverse gene pool and some inherent toughness that is part genetic and partly due to hybrid vigor. Another advantage is that they are a medium-sized breed and, provided they are properly fed, they do not need a huge amount of forage to produce high-quality milk. They have been developed with this aim. In four years, from the first importations and crossings using semen, a really beautiful, compact red-colored cow with a good vessel and conformation to match has been evolved. Their milk is comparing very favorably with established milk breeds like Friesians and Jerseys. This is a breed whose genetics are known and used worldwide nowadays.

Ayrshires

This old breed, which is one of the bases for the Australian Red, was generally horned in the early part of the century. Now it is usually either polled or dehorned and consequently much easier to manage. They have been established as an excellent high-producing cow for a long time. Bulls in the United Kingdom were regarded as being somewhat untrustworthy, the worst after Jerseys. Between World War I and World War II there were a great many fine herds of milking Ayrshires (mainly horned). The farm on which I grew up milked very fine horned Ayrshires.

Dairy Shorthorn

Usually, but not always, a polled shorthorn. They are, as mentioned elsewhere, one of the oldest breeds. There are few more impressive sights than a big, brindled, shorthorn cow; this particular color was much prized both as a family milker and on dairy farms. Many farmers believed it to be superior genetically to the other colors. Shorthorns have been, and still are, regarded very highly on the dairy scene. They are used as crosses when upgrading and establishing new breeds like the Australian Reds, Droughtmaster, etc. They make a really good beef cross when mated out to a meat animal.

Friesian — Dutch, British and American

Mention dairy cattle to the general public and most of them think of black and white Friesians. They are, without a doubt, the most widespread dairy cow in Australia, if not the world. Both their size and production are impressive, but careful breeding and supplies of new genes have been needed to improve their viability and the quality of the milk in the past. The best herds in Australia, which were originally basically British or Dutch Friesians, nowadays often have big infusions of American blood. This has improved their viability and resistance to disease. The old Dutch-based Friesian breed were in danger of becoming too pure. One factor enabled a little new blood to be bred in without changing the color of the animal — to crossbreed with Limousin. This exercise was a success and had become quite necessary at one stage.

Guernsey

This breed emanates from the Island of Guernsey in the French/English Channel and is not very much seen here. There are always a few represented at shows and they have their adherents in the dairy scene. Their particular claim to fame was their high butterfat, like the Jerseys, and therefore rather rich milk. For this reason, both Guernseys and Jerseys are still a popular cross with some of the old Friesian lines whose butterfat was not as high as it might have been. Sometimes a proportion of Jerseys or Guernseys are kept in a Friesian herd just to help balance the lower butterfat content of the milk.

Jersey

This is possibly the most popular Channel Island breed in the world. They are famed for high butterfat and a quite astonishing amount of milk for their size. They are certainly thin coated and skinned compared with many other breeds, and for that reason are often rugged in colder weather. This gives an impression (possibly erroneous) of their being rather fragile. They are not as tractable as they might be. At shows in the United Kingdom when I was young, Jersey bulls were the only ones I ever remember seeing with two handlers.

Meat breeds

Aberdeen Angus

These are usually referred to as Angus or black polls. As their everyday name suggests, they are a polled breed. This is one of the longest established breeds in this country, and they seem to have adapted to Australian conditions in a remarkable manner. They are not a big animal, and this is possibly the reason why they have done so well here initially. Their mineral needs (except for copper) are possibly not as great as those of the big breeds. They are tough, hardy and extremely independent, treating poor fencing — and even good fencing on occasion — with complete disdain. They have been used to "make" several other breeds, like Brangus, and have produced at least one by accident, the Murray Greys. Probably due to their ubiquity as well as their toughness, they were considered good breeding-up material. However, if those who farm them do not recognize their elevated copper needs, due to their all black color, they will be no hardier than any other breed — probably less so. They do have an undeserved reputation for being prone to Johne's Disease, which has been linked with copper deficiencies. Because they are black, their copper needs must be met.

An experiment was conducted by the New South Wales Department of Primary Industries (DPI) to breed what would hopefully become known as low-line and high-line cattle using the Angus breed. The smallest and the largest were selected over a number of years and then bred back until a fairly uniform size was obtained in each category. It was hoped that these breed offshoots might show some characteristics which would be improvements on the basic breed. Apparently this has not been so; they sometimes proved to be slower growers and poorer food converters, so the experiment has not proved really successful. However, they still have their adherents and there are quite a few low and high-line cattle around, especially at shows. I have included this account as many people ask me where they have come from.

Belgian Blue

This is one of the newer breeds in this country and is becoming accepted and established. Their muscling gives them an appearance bordering on the grotesque, but also provides a great deal of extra lean meat. Unfortunately, the breeding necessary to produce this effect has exposed a few genetic defects which are still causing problems and will hopefully eventually be bred out. For this reason they are better used to produce good, first-cross bulls for breeding beef animals. Pure breeding of Belgian Blues is still very much an occupation for the expert cattleman. Many of the stud owners are veterinary surgeons.

Belted Galloway

This is a tough, middle-sized meat animal whose distinctive white "belt" makes it very easy to recognize. They now seem to be on the increase in Australia. They were, of course, originally a native of the British Isles.

Bison

Bison, often mistakenly called buffalo, were imported to this country from Canada and will presumably become another beef producer here once they become established. One of the reasons for the interest in this breed is that they came somewhere near to extinction in the "wild" west of North America where the big herds were shot out for food and leather. Also, they were not considered an adjunct to more concentrated farming. Perhaps the herds will catch on here in the north of Australia.

Blonde d'Aquitaine

This animal is another French breed whose lean beef is one of its chief attractions. Again, most of them are still in the hands of specialist breeders, but they are becoming part of the beef scene now. They will possibly be used for crossbreeding programs.

Boran

These are African cattle of Kenyan origin named for the Borana people of Ethiopia. The semen became available here from quarantine during the early 1990s. They are pure Bos Indicus, which belong to the African Zebu breed. Borans look very like Brahmans, but are apparently unrelated. Like Brahmans, they are a popular breed with ranchers in Kenya. They will be useful to cross into the Brahman breed when genetic diversity is needed.

Bos Indicus

Another African animal that is compact in size and very hardy. These have been in Australia for a few years under quarantine and in the middle 1990s were being released out of quarantine. Their role may well be as a cross with the local breeds.

Braford

This breed, as the name suggests, is made up of from crossing Brahmans and Herefords. The resultant animals are very fine, big beasts and popular with producers. They are used quite widely in feedlots due to their size. They are also found on bigger spreads in the north of Australia.

Brahman

These are well established in Australia and are distinguished by their dewlaps, humps and their black skin which is impervious to the worst that the Australian sun can do. Brahmans are often run as purebreds and are invaluable in breeding the crosses already mentioned above. They have a reputation for tick resistance, which is supposedly passed on to their crosses. They are certainly more resistant to ticks than the English breeds, but breeders tell me that this depends, as with other animals, on the amount of sulfur in their pastures. They are by no means totally immune to this scourge. Ticks are always worst on poor, unmineralized pastures that are low in the lime minerals.

The bulls are extremely popular on the rodeo circuit as they are quite diabolical buckers. The names of several of them — like the renowned Chainsaw — are immortal.

Brangus

Another Brahman cross, this time with Aberdeen Angus, which also has its adherents. One of the reasons for using Brahmans as a cross was to breed a type of cattle which could cope with the harsher conditions in the north of this continent. As with the Brafords, the experiment seems to have been most successful.

Charolais

These French Cattle are famed for their lean beef and, like many of the new breeds, probably owe their presence here to that trait. Now that their stock numbers in Australia are growing, the original troubles breeders had with them, mainly due to a small genetic pool, have been largely overcome. They are a really good cross onto almost any other breed of cattle and are much sought after as their calves mature early. This shows up especially when they are crossed with British breeds.

Chianina

This breed, which emanates from Italy, is one of the late 20th century imports here. It is possibly still suffering from early start-up troubles due to a small genetic pool. They are a very tall, long-legged breed and possibly not as suited to harsher Australian conditions as some of the breeds that are already established here.

Devon (Red Poll)

This breed originated in the United Kingdom and was advertised as being an excellent cross with which to improve other breeds. In particular, they are used to breed out horns as the hornless gene was apparently dominant. They are mainly a beef breed nowadays, but were originally dual purpose. Being another of the red cattle, they have been used in the establishment of the Australian Red.

They are a tough native of southwestern England where they do well on rather light, marginal country. They seem to be well established here and have always been renowned as a docile, hardy and versatile breed.

Droughtmaster

This breed is made up from a Shorthorn/Brahman cross. Originally, Shorthorns were mostly used for the beef herds in Queensland, but they were highly susceptible to ticks. The Brahman cross reduced this problem to an acceptable level, (see section on Brahmans) and the resulting animal is a very fine looking beast. This is another breed that is used in feedlots quite extensively up north.

Hereford

This is one of the older cattle breeds which has been widely used for producing new breeds. This is due, in some part, to the inherent problem with Herefords in this country; the white area that surrounds their eye being responsible for much eye cancer. Crossing out to dark-skinned breeds stopped this unfortunate effect. There has also been, as mentioned in Chapter 5, a tendency to arthritis in the breed. These problems seem to be under control; although the white eye problem with its unfortunate results is still occasionally seen when the cattle are farmed in hot areas. In the southern states, Herefords have been at the forefront on the commercial beef scene and widely used in feedlots. It recently has been suggested that their popularity in this field may be waning.

Highland

Reckoned to be the hardiest and shaggiest of the British breeds, these originally came from the wilder parts of Scotland, Ireland and the surrounding islands. It is believed they were bred up from the ancient Irish Kerry cattle. Some were brought to this country in the 19th century by Scottish settlers. More recently, a bull and two unrelated cows were imported in 1954 by a farmer in South Australia. Since then a large number have been bred up by embryo transfer and artificial insemination (AI), using Shorthorns, Angus and Hereford dams. There are hopefully now enough cattle here to keep the breed viable. They are an extremely tough breed which thrive in mountain conditions. They are on the small side, and their long hair provides good insulation against both heat and cold. It was, of course, the latter factor that made them so successful as the native breed of Scotland.

Limousin

A French breed of cattle now becoming quite widespread here. They are mostly bred up from Shorthorns and were, in their turn, one of the breeds used in Germany to fix Gelbviehs. They are extremely useful as a cross with practically all breeds of cattle mainly due to their lack of dominance, meaning they merge well into other breeds (see above regarding Friesians).

Maine-Anjou

This welcome addition to the beef pool comes from the valleys of the same name in central France. They came to Australia in 1973 and are popular with cattle breeders for their ability to promote high beef returns. Their good milking ability ensures that veal and young stock grow very quickly and, like Limousins, there is a recessive color gene. This allows them to be crossed with other breeds which do not then lose their dominant color. They are popular with both the Japanese market and butchers here in Australia because of their even, well-muscled carcass that has a light covering of fat.

Mashona

These are a smaller Bos Indicus breed of the Sanga type, which is of Zimbabwean origin. They are a compact and very active foraging breed that would be more suited to range management than being contained in paddocks. Their semen and some animals are becoming available in Australia now.

Murray Grey

A breed of cattle which are peculiarly Australian. It originated at the beginning of the 20th century on a property in the Murray River area on the New South Wales/Victorian border. The first ones were a "sport", probably a recessive, that suddenly appeared in a herd of Angus cattle. The quite different coloration was originally a source of some embarrassment to the breeders. They certainly were not expected to found a breed, but as often happens with this sort of genetic "accident", the new breed was quite different in type from its Angus forebears.

Farmers were quick to realize the advantages of animals that were quick developers and good feed converters and their marbled meat is now an additional bonus in the export market. They vary in color from silver to dark grey and are noted for their docility. The black skin, which they inherit from the Angus, means that they are an ideal animal for hot and harsh Australian conditions. In the early days, a lack of genetic diversity in the breed led to calving and viability problems. These have now been overcome; the breed plays a prominent part in the local scene and is popular for use as a cross with most other breeds of cattle.

Piedmontese

These cattle, emanating from Italy, were bred up originally from Zebu Cattle (Bos Indicus) from Africa over Aurochs cattle — the oldest base of many cattle breeds in the world. They are a comparatively new arrival in Australia, and are reckoned to be particularly useful as a crossbreed over dairy cattle to produce a double-muscled, quickly-growing veal calf.

Salers

A compact French breed, they are largely red in color and known for their docile and friendly temperaments. They are a reasonably small breed that produces good meat. Because of their good manners, they are popular with newer and inexperienced farmers.

Texas Longhorn

The original United States ranchers' cattle; these are descended from the cattle brought over and farmed for many years by the Spaniards. Their rapid spread (and decline) is touched upon in Chapter 1. They seem, in spite of their large horns, to be tractable and popular here.

Tuli

Another African breed which, like the Boran, is newly available in Australia. It is basically of pure African Sanga breed and is unrelated to Bos Indicus or Bos Taurus. It descends from animals that have been in Africa for 5,000 years. They have been selected and improved in Botswana and Zimbabwe, and are considered a very adaptable type of cattle which do well in semiarid conditions. They and/or their semen are now available here. They could be a useful breed both genetically and for range conditions in some of the drier parts of the world.

Water Buffalo

In the middle of the 19th century, these were brought to Australia chiefly as a beasts of burden. However, they adapted to the northern parts of Australia with great speed once they were let loose and nearly reached plague proportions at one time. Now they are much in demand in the "third world" whose own stocks have become very low, and are a regular export. (Australia's freedom from some of the more unpleasant cattle diseases makes it an excellent country for cattle exports.)

Wagyu

This breed evolved in Japan in the first half of the 20th century by crossing Mishima cattle and some European breeds. They are a breed which has hitherto been unobtainable from Japan as the Japanese preferred not to export it. However, in 1975, four bulls were almost literally spirited out of Japan to the United States by a local businessman and a Texas rancher. They did not prove very popular in Texas due to their light-weight appearance, and changed hands several times.

Almost too late, their export potential was realized. The Japanese are very particular about the kind of beef they like, and many of the European breeds find little favor with them. However, the Wagyu cross provides what they want — Kobe-type beef. There are stories of semen thrown away and animals nearly given away before their value was realized. They are now well established in Australia and have been mentioned in Chapter 1.

Dual-purpose breeds

Brown Swiss

These are a smaller-framed animal than the usual beef breeds in Australia and, therefore, much suited to the lesser acreages. A hallmark of the breed, which are often bred up from shorthorns here, is the white nose surrounds. In its homeland of Switzerland, this breed is regarded as dual purpose. One of its much-prized traits is its docility in both sexes which gives great ease of management. Their numbers in Australia are still fairly low, but their adherents are growing and they are certainly set to take their place in the beef and dairy scene. Their milk is well regarded for having a high protein-to-butterfat ratio.

Dexter

This pint-sized (52-inches high) and very tough little breed of black cattle is usually found on smaller set-ups. Milking it on a large scale is difficult possibly because of the shortness of its legs, but it is a very productive and equable beast that is much prized by its adherents. Due to its docile temperament, it is eminently suited to small-scale farming where many of them are in studs. The average weight of the breed is around 400 pounds, which belies its small size — due, as mentioend, to its rather short legs. It is popular in the beef market and as a house milker and, unlike other small, made-up breeds, is an entity in its own right.

Gelbvieh

A dual purpose native of Germany, the name literally means "yellow beef" and they are sometimes known as golden cattle. The breed was formed in Germany using Simmental, Shorthorn and Landrace cattle, their docility and ease of handling being a characteristic. They are prized as early breeders who produce their quick-growing progeny very easily. Many beef breeders have found that a final cross of Gelbvieh onto a Brahman/Shorthorn cross produces a top animal with great market potential.

Illawarra Shorthorn

These are, as the name suggests, an Australian breed — initially accidentally evolved from Shorthorns and Ayrshires. They have been used largely as one of the base breeds for Australian Reds and, once they had got over the scarcity of numbers in their early days, they have become known as a useful dual purpose-animal.

Pinzgauer

These striking-looking cattle, with their white upper body and another stripe that goes along the belly, around the chest and onto the tops of the legs, is regarded as being a good dual purpose animal. It is a late 20th century newcomer on the Australian cattle scene, which will help ensure that lack of genetic diversity never occurs here again.

Chapter 3

Improving Farm Efficiency

Setting up a farm

This is a matter of making the farm as easy to run as possible. If the whole operation is not to be waste of time, the management will include remineralization and land improvement (see Chapter 4). If the farm is being converted to beef or dairy, or has just been acquired, a study of P. A. Yeomans' techniques as described in *Water for Every Farm* might be helpful, especially if the farm is in an undulating area.

Fencing and gates

Make fencing a one-time expense and do a really good job the first time. It is no economy to put up bad fencing which will need replacing down the road. Cattle can be effectively controlled without masses of barbed wire, but the fences must be top quality. Remember that cattle always tend to go *in front* of the wind, unlike sheep and goats who go towards it. Therefore, set up your fences with the prevailing wind in mind.

It is also better to have slightly more gates than you think you may need. This is due to the fact that it is difficult to impossible to herd one bovine escapee, and the usual practice is to take the mob so it joins them. It is annoying and time consuming to have to backtrack the mob to a distant gate when an additional one on the other corner of the paddock would have saved much handling. Most farms with welding equipment make their own gates; two-inch pipe frames make really good, long-lasting gates and are not easily bent. I have had my share of removing gates and hammering them flat because they were not made of strong enough material. This is a great waste of time, and even worse if the cattle break through one.

Water — conservation and irrigation

In most places there is plenty of rain. It often falls at the wrong times and in too great amounts to store. Yeomans' system of dams that start on the high ground means that the irrigation is by gravity; therefore, it is cheap and highly effective. It also means that every farm *is* capable of catching and storing its own water. Since one may ultimately have to do so, it is best to be prepared beforehand.

Keyline, the name Yeomans chose for this type of farming, is eminently suited to undulating or hill country. Dams that reticulate the water around the farm by gravity are used and are fully described in the book. The use of soil aerators for improving and revitalizing the soil are also part of the system. The initial setup takes time, but so does any irrigation system on the farm, and the result is well worth the effort. The farm mentioned below has a problem with blue algae, it is located in southeastern Australia and is irrigated and run entirely on Keyline principles. Ron and Bev Smith have been selling their organic milk for quite a while now, and a visit to their farm shows the value of Yeomans' books (see Selected Reading).

A good strategy for containing rainfall is a fertile, permeable soil. One of the best tools for achieving maximum saturation when it rains is an aerator. The catastrophic floods in northern Victoria and in the Missippi basin of the United States in 1994 were made a great deal worse by the sadly compacted land which absorbed little, if any, of the water that fell. In a healthy, friable soil, the first inch of water that falls should be fully absorbed before any run-off starts. As will be seen in the next chapter, compacted land can be due to incorrect calcium and magnesium ratios in the soil.

I own a Wallace aerator (similar to a Yeomans) and have no hesitation in saying that it is equal to a half a ton of neutralizing material to the hectare (about two and one-half acres) by its action in sweetening up sour, low-pH ground. P. A. Yeomans, who made a similar implement, may have gotten the idea when he was working in United States from their conservation department. According to some papers I have called *Readings in the History of the Soil Conservation Service*, collated by Douglas Helms, an implement very similar to Yeomans' plough has been in use by that department since before World War II. Whoever is responsible, an aerator is one of the most valuable aids to soil regeneration that we have.

The aerators are equally useful when paddocks are irrigated. When I had an irrigation farm, I took less runs of water than any other farm in the district. I usually found three waterings a year were adequate, even when everyone else was letting the water through once a week during the summer months. Because the land had been remineralized and was no longer compacted, and the water was going *into* the soil instead of *over* it, the effect of the water was maximized. Water from irrigation systems is expensive and even one's own water costs money to conserve, so letting it run over the land without being absorbed is wasteful and causes mineral losses by leaching.

Salinated land has become the great issue as we go into the new millennium and in many areas it is largely due to too much water being used, or rather, wasted. Maximizing the effect of every ounce of water used will help to prevent and finally solve this problem in areas like Victoria where it is largely caused by misuse of water. Salinated land also responds to top-dressing with gypsum. In the United States it was found that as little as a quarter ton to the acre went a long way toward sweetening up the soil.

Stress areas and dam management

Cattle always pug up their moving areas and around their water. Dairy cattle have to be brought in twice a day for milking and care; they obviously place their pathways under a great amount of stress. Concrete, where practicable, is always useful. It is possibly a better idea than gravel made from large stones which quite often causes foot injuries. All paddocks should open onto pathways that lead to the milking complex. In this way, the minimum area of land is rendered unfit for grazing.

Water should be available in troughs surrounded, if possible, by concrete aprons. It is *not* good practice for cattle to drink from dams unless on range conditions. Even then, the damage their feet do to the dam approaches, along with the effects of the dung being dropped in the dam, is bad farming practice. If possible, dams should be fenced off and the water pulled up by a windmill into a tank and reticulated to troughs. An advantage of this method is that trace minerals, licks and seaweed products can be quite easily added to water, or the lick stations can be put by the troughs if supplementation is needed.

Another disadvantage of letting cattle have access to dams is blue-green algae, which tends to take over when the manure builds up in the dam. This is particularly a problem if there are no rushes or plants around the dam to use the extra phosphates.

In Western Australia, where the dams tend to have large, open fronts for the water to run in, rather than leading it in from channels as is often done in the eastern states, the risk of the algae is even greater. When it rains heavily, the manure from the paddock can be washed straight into the dam.

One farmer near Kojonup (Western Australia) had an excellent idea for stopping this. He built A-shaped batters of loose rocks about three-feet high and six-feet in width right across the mouth of the dam. His property, like many others, had a rock problem and this way he put the rocks to the best use. It did not stop the water running in, but it effectively filtered out the dung.

Blue-green algae cause irreversible damage and the animal dies from total phosphate starvation caused by the algae. This happened on Ron and Bev Smith's organic dairy farm in Gippsland, Australia. In this case, the algal bloom was triggered by the effluent seeping from a piggery not on their farm, but from a farm above theirs. No one had realized that it was reaching the dam until it was too late. Blue-green algae is fairly easy to see — there is a shadow of that color on the water. Putting dolomite and copper sulphate into a dam will help control it, but six days should be allowed before stock is re-introduced to the dam.

Prevention, in the form of plant life in the dam, should be considered. It is normal for plants to grow in water and in my youth all the lakes and dams had water lilies in them naturally. Ultimately, it is best to keep as much manure as possible out of the water supply for obvious reasons.

Supplying minerals on range or in feed

Beef cattle in range conditions are nearly always watered from artesian bores and troughs. As mentioned previously, these provide effective gathering spots when feeding extra minerals in the form of licks which can be left out by the bore or water source. Artesian bores have another advantage in that many of them are quite high in minerals including sulfur. All cattle need supplementation with minerals at some time and when conditions are hard, like those in the recurring droughts, they may make the difference between survival and death.

It is comparatively easy to feed additional minerals to dairy cattle that are bail fed. The lick is fed as part of the ration as well as being available on demand. That way, they are always supplemented.

Milking systems, crushes and yards

It is essential to install the best milking systems and handling complexes for cattle that you can afford. There is nothing worse than trying to draft cattle with inadequate facilities. Jervis Hayes, a vet from Adelong, Australia, commented in the press: "Cattle crushes are either made by cockies (cow farmers) who know nothing of engineering, or engineers who know nothing about cattle." I fear Dr. Hayes had been unfortunate in some of the farms he has had to work on, but I too I have seen some very bad setups and they cost just as much as good ones.

Gates that do not close quickly and effectively; crushes that need two men to move the levers and do not hold the beasts in place, and badly built yards are time wasting if not downright dangerous. Systems for drafting cattle are expensive, but if done properly in the first instance, they should be a one-time capital outlay. Cows are extremely strong and there are no short cuts.

For milking, the choice seems to be between pit herringbone systems and rotaries; each method has its champions. The snag with pit systems, where inorganic compounds are used, is that some of the teat sprays are extremely bad for the operator to breathe. Chlorhexadine-based dips in particular, when micronized, can be lethal. I think that farmers have now realized that the old iodophor sprays are much safer and just as effective.

Cows who receive the right minerals and are in good health do not need teat dipping. I ran a good herd of milking goats and never ever used teat dips. The sphincter at the bottom of the teat stays open for about 20 minutes after milking and for 15 minutes after a calf suckling. A douche of cold water will help it close. I have yet to see any animal go straight out of the milking parlor and lie down. Flies could be a source of trouble, but I never found they were, even in districts where they generally caused problems.

All that should be necessary at milking time is to brush the udder free of loose dirt so that it cannot fall into the inflation, and then do not allow the first squirt of milk to go into whatever sort of milk collection system you are using. This way the plug at the end of the teat does not go into the vat. Doing this can make the difference between clean milk and an *E. coli* bacteria count of 20,000.

Cattle management and routine

Cattle, like all animals, are great on routine and if they are properly handled and moved around regularly they very soon learn the ropes. I know several farmers who use cell grazing and they say that after the first two or three weeks, the gates only have to be opened and the cattle automatically walk into the fresh pasture. Dairy cattle go to and from their paddocks on cue, especially if they know they are to be fed when they come in.

Years ago, I offered to mind a jersey cow for friends going on holiday. Someone suggested I was mad, since watching the entire family trying to get the cow in for morning and evening milking was one of the sights of the village. The first morning the cow viewed me with some caution, not believing the bucket really held a bit of food.

After that morning, there was no stopping her — she rocketed into the makeshift bail the moment I came out and called her.

When buying cattle or starting on a new mob, find out how the previous owners managed them — on foot, horse or bike. Cattle very soon become used to one mode of handling; I have seen mobs that had always been worked with horses, and had never seen a person on foot. The sight of a human walking toward them was enough to make them panic in all directions, yet they were totally amenable when one was on a horse (the same applies to farm bikes). With an unknown mob, failing a good dog, it is wise to carry a stock whip and know how to crack it effectively. Cattle are too large to be allowed to walk all over one. When I first started working with cattle in the United Kingdom, I was left to take delivery of a new mob from market. My father said all I had to do was stand in front of them with my arms out and they would stop. As I picked myself up and felt my wounds, I thought he had apparently always had very tractable cows.

Hay and crop fodder management

Adequate storage for hay is a necessity, as much or even more so on irrigation than on dryland farms. An irrigation farmer whom I questioned about his many empty hay sheds at the beginning of the worst drought in recent history in Australia (1983-1984), told me he did not think he needed hay — he had irrigation. By then his cattle were dying of starvation as the irrigation water was no longer available.

Even cattle on irrigation need a certain percentage of dry matter to maintain the health of the gut. The twice daily bail feed is *not* enough to provide this. Hay, preferably a good grass/clover mix grown on remineralized soil, fills the bill admirably. The advantage of this kind of hay is that it will keep in the shed for several years and still be fit to use, unlike grain hay. The paddocks from which the hay is grown should be alternated, so that the cattle dung plays its part in providing the natural fertilizer along with lime, dolomite or gypsum, whichever is indicated on the soil audit.

In other countries and occasionally here, turnips and mangel beets (in the colder climates) are grown for winter feeding. They are fed either directly from the paddock or when cattle are confined. The amount of rainfall in European countries usually dictates that large numbers of heavy animals be kept off the paddocks in winter, otherwise they would totally ruin the soil structure. In the northern United States and Canada, frozen winter conditions make the same kind of management a necessity, and roots provide a good addition to the normal diet of concentrates and hay.

For spring and summer, fodder kale, rape and occasionally grain crops are grown for cattle to eat standing. The cattle are usually controlled by strip electric systems or movable fencing of some kind. In Germany and Japan, Russian comfrey, a plant very high in natural minerals, is largely used as summer feed for dairy cattle and is reputed to produce good milk yields. Comfrey, like aloe vera, is one of the few plants that is high in natural vitamin B12.

Utilizing manure and slurry

The slurry from the dairy complex is one of the most valuable tools for maintaining the fertility of the farm and can be used either for the cropping and hay-growing operations or for top-dressing the grazing paddocks as they are rested. In my opinion, any fresh manure is better recycled through a growing crop rather than put directly on the grazing paddocks. Mixing the slurry with rice hulls, sawdust or other cheaply obtainable organic matter and letting it compost for a few months is the ideal and will help equalize the pH. Adding some lime or dolomite, about a bag to the ton or per 1,000 liters (264.2 gallons) of slurry, is also a good strategy.

I went to a field day on a Gippsland dairy farm that had been addressing the problems of poor animal health, including a very bad fertility record. Prior to changing farming methods, artificial fertilizers had been used heavily and, in the process, tied up magnesium, phosphorus, trace minerals, and in particular copper, as well as upsetting the soil ecosystems.

It had taken seven years to bring the farm back to a balanced level — far too long. The farmer's refusal to use the slurry had caused the delay. You cannot farm organically by default, which means you stop using artificials but make no effort to replace them in the farming operation. Superphosphate came into use to replace animal manure fertilizers, which had been used for generations to maintain the health of the land in the northern hemisphere. It is easier to spread a powder than manure and it is often easier to obtain bags of superphosphate than loads of manure. Now, we have composted manure available from feedlots and, if within a viable distance, it is well worth considering. It has to be composted for two years and passed by a certifying body as being free of heavy metals and drugs.

Dairy farms, on the other hand, have their own supply of ready-made organic manure, spread by what my youngest son always referred to as walking manure spreaders — cows. If the soil is regularly analyzed and remineralized, the farm will go from strength to strength at the same time. However, if manure from the sheds or dairy is incorrectly used, it can cause nitrate buildup and become extremely dangerous to the local groundwater and ecosystem.

Fletcher Sims and Malcolm Beck have both made composting an art. Each has written books on the subject detailing how they operate and the machinery used. Some machinery has been invented especially for the task. Acres U.S.A. in Austin, Texas, is a great source for books on agricultural subjects.

Chapter 4

Using Soil Analysis to Improve Your Farm and Stock

This chapter gives examples of three different soil analyses and comments on each. For more exact information on the minerals mentioned here, consult Chapter 8 which deals in detail with each one. If you are having a soil audit done, your consultant will tell you how the soil sample should be taken. The following are the basics.

To take a soil analysis, be governed by the treatment the farm has had rather than the different soil types. For instance, if artificial fertilizers have been used on one half of the farm and not the other, or one half has been cropped and not the other, take two samples of about 30 sample cores from each part. An apple corer is one of best implements for this job, it is all stainless steel and the right length. If the soil is soft enough, a piece of plastic pipe, three inches for pasture and four inches for arable land, is usually recommended. Mix the core samples in a non-ferrous (i.e., plastic) bucket and send a pound of the mixture to the agency for the analysis. Using old bits of stainless steel pipe (usually ex-dairy) to take core samples is *not* a good idea, they are often neither stainless nor steel, and the result is contaminated readings of trace minerals which is of no help at all.

What is a soil analysis?

In the United States, an analysis is often referred to as a soil audit, which is as good a title as any. As with any audit, you have to know the *whole* situation. Anyone who is serious about the health of the land and those who live off it will concede that the necessity for a soil analysis is indisputable. This must be done by an independent firm, who, in the words Dr. Geoff Johnson of South Australia, "Sells advice and not products." We have at present one such agency in Australia, SWEP; there are also many in the United States. Each soil testing operation will tell you exactly how to take and handle soil samples.

Ascertaining the state and contents of your soils is the single most useful operation you can do on the farm. On it depends the health of your crops, pasture and stock as well as your own if you are growing your own food or handling harmful chemicals. The visual aspect of a farm will tell very little in many cases, but the quality and quantity of the produce and the health of the stock will tell it all. Many so-called "pasture boosters" are usually of phosphatic and nitrogenous origin and make pastures look very good to the uninitiated. The animals on them do *not* look good.

In any country remineralization is always the first step and, according to Neal Kinsey, one of the best in this field now that Dr. William Albrecht is no longer with us, it is just as important worldwide, even if mineral deficiencies are not as inherently poor as they are in Australia. Only after remineralization can the farmer start to build up the fertility of the soil as the mycorrhizae and other microbes swing into action. If the calcium/magnesium levels and ratios are incorrect, nothing can work. In this respect, I would suggest anyone who is really serious about this read the works of Dr. William A. Albrecht, previously mentioned, as well as *Hands-On Agronomy*, by Neal Kinsey and Charles Walters. If you get a chance to hear Neal Kinsey speak, or can attend one of his workshops, make every effort to do so. (The above-mentioned books are available from Acres U.S.A., who are the publishers.)

The following soil analyses are included as examples of what to expect from the company undertaking the analysis and what to look for. I should explain that the pH in water is about .04 percent higher than when taken from the soil calcium. This fact has been used for years by fertilizer companies of little conscience when trying to persuade people to put on still more superphosphate.

Plant nutrients exist in the soil in three forms: unavailable, exchangeable or partly available, soluble or readily available. For the purposes of interpreting the analyses, the exchangeable nutrients exist as ions bound to the surface of the colloidal particles of clay and organic matter. These particles are negatively charged and can attract and hold positively charged ions known cations. Cations are formed by calcium, magnesium, hydrogen, sodium, potassium, zinc, copper, manganese and iron, which are held in the soil in exchangeable form and thus are available to plants.

The CEC (Cation Exchange Capacity) is a measure of the exchangeable conductivity of the soil being analyzed, without which the soils with a perfect pH and everything else would not function. Very weak land at Esperance (coastal Western Australia) shows a CEC of six or seven, whereas an analysis of virgin rainforest showed a CEC of 51.59. The different desirable level of nutrients shown in the charts on the following pages are due to differing CEC levels. The code NR in an analysis stands for not required.

Anyone desiring a more in-depth knowledge of soil chemistry is referred to Ted Mikhail's papers in the bibliography or to Dr. William A. Albrecht's collected works.

Soil analysis A

This is a general sample taken fron a farm in Southern Victoria and provides a fairly usual picture of an exhausted soil that is mitigated by good organic matter and a "bank" of locked up phosphorus. Although the pH is low, the addition of the required lime minerals, including gypsum, if necessary, to raise and balance the all-important calcium/magnesium/sulfur levels, followed by careful use of an

Soil Analysis A

Item	Result	Desirable level
Color: dark grey		
Texture: fine sand loam		
pH (1.5 water)	5.00	
pH (1:5 01M c12)	4.5	
Electric conductivity, EC us/cm	200	< 300
Total soluble salt, TSS ppm	660	< 300
Available calcium, Ca ppm	860	1800-2400
Available magnesium, Mg ppm	160	430-600
Available sodium, Na ppm	207	< 184
Available hydrogen, H ppm	114	< 64
Available nitrogen, N ppm	8.20	25
Available phosphorus, P ppm	13.60	30
Available potassium, K ppm	121	190
Available sulfur, S ppm	4.10	7
Available copper, Cu ppm	0.20	2
Available zinc, Zn ppm	3.00	7
Available iron, Fe ppm	23	> 20
Available manganese, Mn ppm	2	> 20
Available cobalt, Co ppm	0.03	> 1.0
Available molybdenum, Mo ppm	0.30	1.0
Available boron, B ppm	0.60	0.6-1.0
Total Organic Matter, OM %	9.90	6-10
Total Phosphorus, TPN ppm	217	
Cation Exchange Capacity, CEC	16.04	
Exch. Sodium Percentage, ESP	4.93	< 5
Calcium/magnesium Ratio, Ca/Mg	3.23	2-4

Percentages of Ca, Mg, Na, K and Percentage of CEC

	Percentage	Percentage of CEC
Exchangeable Calcium, Ca	62.90%	23.57%
Exchangeable Magnesium, Mg	19.47%	7.29%
Exchangeable Sodium, Na	13.14%	4.93%
Exchangeable Potassium, K	4.49%	1.68%
Exchangeable Hydrogen, H		62.53%
Total:	100.00%	100.00%

Recommendations:

Gypsum	1 ton to 2.5 acres
Dolomite	1 ton to 2.5 acres
Lime minerals	

aerator, should start it on the road to recovery within a year or two. Compaction, whether induced by unbalanced lime minerals, heavy machinery or incorrect farming practices, cannot really be dealt with unless the main minerals are brought into line first. Then, and only then, can the farmer cut hay off the farm secure in the knowledge that it will do his stock active good. In the words of a farmer I know in Western Australia, "Dad told me never to let hay leave the farm unless it was inside a cow." Unfortunately, selling large amounts of hay off a farm is the most difficult kind of farming. The law of returns dictates that what is taken off must be put back or the farm will degenerate.

In four or five years or less with top-dressing (possibly only once) and attention to soil management, a grazing property should reach a good level of sustainability.

The low calcium, magnesium and sulfur are the limiting factors on this property. These levels, accompanied by a high hydrogen, point to a weedy, rather sour, soil, (see below). The gypsum (calcium sulphate) is needed to raise the sulfur to a better level. The stock may have lice and poor digestive performance until this is done.

Topdress as suggested and follow with the use of an aerator as soon as the ground is suitable — about four inches of rain is needed to use an aerator without doing too much damage to the land — and the improved health of the stock will be seen quite soon. The low available phosphorus will start to rise as the lime minerals are applied.

Neal Kinsey mentions cases where the available phosphorus has doubled in the first year once the calcium was brought up to the required level. He also mentions that adequate rainfall is needed. It will possibly be about 18 months before any noticeable change to the grass types and weeds will be seen, but before that there will be a noticeable improvement in the cattle and stock. In the United Kingdom, the recovery of the soil took five months on one farm. That is the difference between a soil that is basically good quality and one as poor as what is usually found in Australia.

Provided the aeration is carried out regularly each year (when possible), the farm should be fairly well recovered with a pH around 6.0 in four to five years. I have seen this happen on our own farms, all of which started very sick, low in essential minerals and pH to match. It can happen more quickly, but it is better to be prepared for the full time. After that, it will be a matter of maintenance, checking the lime mineral balance and keeping the organic matter in good heart.

If hay is being sold off the farm, remineralization will be a fairly steady necessity every two or three years, accompanied by some form of organic matter. The law of returns, mentioned before, has to be observed; otherwise the farm will be steadily depleted. Stock puts back its own manure and the system can be more or less sustainable after a time. Commercial hay producers and grain farmers take more off their land than those who pasture animals. Those farmers that do not burn off but incorporate their residues stop a great deal of this waste. If the hay is eaten on the farm during the winter months, as is generally done in the Northern hemisphere, this wastage does not arise. In Australia, and probably the warmer parts of the United States, it can and does. The time will come when large-scale composting will have to be considered, as well as fairly continual monitoring of the soil minerals.

Organic matter

The reasonable organic matter is the saving grace here, because everything that is applied will be held in the ground and fully utilized by the soil organisms. Had the organic matter been low, the stock would already have been affected by varying conditions of ill health. If the organic matter shows much above the desired level, it will effect the pH levels. So in part, a low pH will not be as serious as it looks. A low pH can mean just that, but if the organic matter is high, much of the low pH is due to its actual acidity. Organic matter is acid, so it is necessary to know the whole picture.

I do not consider it advisable to spread more than a maximum of one and a half tons to the acre of lime minerals in a given year when remineralizing. My feeling is that the soil can only absorb a certain amount each year.

Low pH

At a talk a short while ago, one farmer told me he did not need to take a soil analyses because he had a pH meter and was constantly monitoring it. Nothing could be further from the truth. You can have a pH in the desired range, 6.5 to 7.0, and have the whole farm badly out of balance. High potassium and sodium can produce this effect and, as the calcium and magnesium are way below what they should be, the farm will be in a terrible mess. The lime minerals *must* be balanced first. Neal Kinsey suggests that 60 or 70 percent of calcium and 10 to 20 percent of magnesium will be acceptable in heavy soils, but in light soils 70 percent calcium to 10 percent magnesium would be better. The equation should add up to about 80 percent. It is fine to have a pH meter; it gives one a guide *after* remineralization and can be quite an interesting exercise.

In the inexpensive range, I favor the pronged variety pH meter. Its readings seem to be fairly consistent. This type might not do for consulting work, but is perfectly efficient as a monitoring tool when used on your own property. In the higher-priced range there are several models available (at around $100).

The worst effects of the low pH on the above, or any, analysis is that the trace minerals are unavailable to stock. All, including selenium (Se) which is not shown, will be in such low supply that animal health will be at risk already. Supplementation with licks is always necessary in Australia with its variable climate and, indeed, in most places in the world, but in cases like the above, it will mean the difference between success and disaster. Until the pH reaches 5.5 and above, it is a waste of time and money to spread trace minerals as they will be lost. Often by the time that desired level is reached, many of the trace minerals are coming back into the food chain anyway.

As will be seen from the previous analysis, pH levels are reckoned in water and calcium; the former is .5 higher than that reckoned with calcium. I have always found working with the lower level produced the best results in the long run.

Potassium

Looking at the previous sample, the potassium level is half what it should be. Provided any breeding stock are supplemented with cider vinegar before calving the first year, all should be well by the second one because, after top-dressing as suggested, potassium should become more plentiful. Iron is in good balance on this farm which is unusual; therefore, it will not interfere with potassium uptake as often happens here when iron is too high. Dystokia (pulling calves) is a sign of low potassium in the fields.

Sodium

The high sodium level is indicative of the general picture and it will come back to somewhere near the norm when the lime minerals are applied. Needless to say, a mineral cannot be materially reduced, but it can be brought into balance by raising the essential ones to their correct level.

As far as stock are concerned it is always a good idea to have rock salt available in all paddocks all the year round, just as the base stock lick advocated in this book should be available at the same time. Occasionally stock go mad for salt, and if we do not have an analysis, I have the farmer get one as fast as possible. This is because an abnormal appetite for salt can mean a bad potassium shortfall, and the animals are not actually after salt at all, but potassium. On one farm where the sheep were eating salt by the kilogram, I had the farmer spray their hay with cider vinegar which contains natural potassium and the sheep left the salt alone from that time forward.

Eating high amounts of salt is dangerous as it can cause instant heart stoppage; this is because salt converts to potassium in the body on occasion and too much potassium causes heart failure. To understand this fully, it is necessary to read Professor Louis Kervran's book, *Biological Transmutations*, which is the history of about 10 years of work, done at the University of Lyons in the 1960s, on the ways in which the transmutation of various elements can occur. Space does not permit further explanation, but this book is readily obtainable from Acres U.S.A. Kervran is the author of 12 other books on similar subjects, but not all of his books have been translated from French.

Phosphorus

Many similar analyses show low phosphorus, often in spite of repeated top-dressing with fertilizers. This one is no exception, but as the total phosphorus readout had been requested, it should be remembered that the 217 parts per million will become available as calcium rises and the soil improves. Phosphorus is only spread on paddocks to replace the non-existent animal manure which was available in the northern hemisphere after the stock had been shut in for the winter. With a "bank" like the one on this farm, and the stock they are running, it is clearly unnecessary to continue N-P-K farming. The soil has to be revitalized and lime minerals used to enable the plants to grow. As the calcium/magnesium levels come back into balance, along with the pH as well, this "bank" of unavailable phosphorus will start to be utilized in the ensuing years.

I have yet to see animals disadvantaged by too low phosphorus; as long as they receive the right minerals, they seem to do just fine. Should it be necessary to apply phosphorus, rock phosphate can be applied on the higher pHs and reactive phosphate (RPR) on the lower ones. However, given the fact that artificial phosphorus ties up copper 100 percent (according to André Voisin) as does artificial nitrogen, I would not feel that either is necessary.

Nitrogen

Most people do not realize that when a soil is in good health, the plants obtain up to 80 percent of their nitrogen from the air and rain; lightning is another potent source. Whether chemical or organic, high-nitrogen fertilizers are not good for pasture, they kick it on too fast and inhibit copper as well. Most animal manures dropped naturally in the paddocks are safe providing remineralizing, aerating and harrowing are undertaken where necessary. But if the manure is obtained in bulk (uncomposted) from feedlots, stables, piggeries or intensive bird operations, it is better if the supply is composted first. There are feedlots that do composting themselves and the finished product has been tested free of all chemicals and drugs.

Pig and chicken manure, especially, should be totally free from foul odors and should be composted for at least a year. Any manure of this nature is best used under a crop or for hay. It is unwise and unsafe to put it out on pasture where stock are grazing, unless they have unlimited access to other land as well.

As the land comes into natural balance, the nitrogen synthesis starts to work again, but in chemical farming, artificial nitrogen is often needed. Nitrogen is just beginning to show up in underground aquifers in Australia. In the United States, this is already a dangerous problem. The use of artificial nitrogen should be stopped before it is too late and all our underground water becomes contaminated.

Trace minerals

It would be a total waste of time and money to apply the missing trace minerals to the soil when the pH is as low as in this analysis. The extremely acid conditions would cause them to be sulfated and washed out of the soil by the first rains after application (I learned this the hard way years ago by listening to ill-informed advice). Often, as the pH climbs back to somewhere near the desired level, these trace minerals become available again. Soil microbes and organisms get into top gear and the interchange between the soils lower down, where the minerals have not been leached or sulfated out, can bring about an almost magical improvement. Soil organisms, from the smallest bacteria to the largest earthworms and dung beetles, do not like compacted and sour soils that go with a low pH and poor lime mineral levels.

General notes on iron, copper and molybdenum levels

This applies to all analyses and for further reading see the sections on those minerals in Chapter 8. These three minerals interact. Without copper, iron cannot be assimilated; high iron unbalances practically everything else; molybdenum inhibits copper, so missing minerals in the soil begin to sound like a whodunnit. Even in the middle 1900s in the United Kingdom the spreading of molybdenum was frowned upon. Because England is not high in iron like Australia, basic slag (a by-product of the iron industry) was spread every five years and to an extent this was reckoned to take care of the molybdenum levels. But we were taught that molybdenum sorts itself out as the basic health of the soil is regained, and in my experience this seems to happen. Iron shortfalls are rare in Australia, although not in Europe.

Years ago, within three months of arriving at my Garfield, Victoria, farm where the iron was initially 1600 times too high, my herd of commercial milking goats showed every sign of imminent death from anemia. Startlingly high amounts of copper sulfate added to the bail feed for 10 days brought them through and, incidentally, stopped the liver fluke infestation as well.

Copper is generally present in reasonable proportions, but it is always inhibited by artificial fertilizers such as superphosphate and urea in any form. It also can be made unavailable due to an out of balance pH.

There is a strange anomaly here. Two so-called weeds, St. John's Wort (*Hypericum perforatum*) and Patterson's Curse (*Echium plantagineum*) are so high in copper that white animals often die of an overdose should they be hungry enough to have to eat them. Black animals, having a higher copper requirement, can tolerate them and thrive. I always imagined that these plants would only grow on land high in copper and could not understand the situation of some friends whose St. John's Wort was all destroyed by a flood, whereupon their goats came down with every copper deficiency disease in the book. I imagined that the surface grass, of which there was plenty, would contain the copper from the ground. Both these two so-called weeds are so deep-rooted that they dredge up their copper from well below the levels where pasture plants would obtain their minerals. This is probably nature's way of bringing back copper to denuded land. I realized this later when staying on a friends farm at Chiltern, where there was plenty of St. John's Wort and the copper in his analysis showed a fifth of what it should have been.

Heliotrope (*Heliotropium amplexicaule*) is also high in copper. On the other hand, it is a shallow-rooted weed that is generally found where copper levels are adequate; however, the pH and lime levels are always too low and out of balance. Heliotrope is the plant that causes stock to die of the yellows (jaundice) when it grows after unseasonable summer rains on land deficient in calcium and magnesium.

Hydrogen

From the point of view of an advisor who has to work on the analysis but has not seen the farm in question, the hydrogen figure (which is usually too high as it is here) in the previous analysis is of particular interest. This figure means that the calcium and magnesium ions on the soil colloid particles are rapidly being lost and replaced by hydrogen. When this happens, neither good grass nor crop will grow and, unless the farm is heavily top-dressed with the necessary lime minerals, the health of the stock and the paddocks will continue to deteriorate. Weeds will become endemic and few, if any, nutritious varieties of grass will be able to survive. Those that do survive will be of low feed value and largely unpalatable to stock. The weeds will cover an enormous range, among them cape weed, sorrel, dock, barley grass, geranium, hoary cress, heliotrope, St. John's Wort, Patterson's curse, onion grass (also called Guildford grass or nut grass), oxalis, fireweed, Crofton weed and many others.

In actual fact, weeds show what is missing in the soil. In *Eco-Farm*, by Charles Walters, there is an excellent chapter showing what deficiencies result in which kind of weeds. It is obvious that sprays are *not* the answer; improving the soil health is what is needed. The same goes for the poorer kinds of grass, Yorkshire fog and bent grass are somewhere near the bottom of the line, aeration and rem-

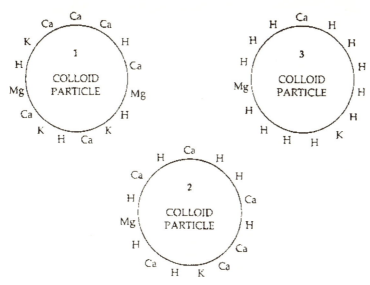

1. Shows a healthy soil particle with adequate minerals. Calcium (C), magnesium (Mg), potassium (K) and hydrogen (H) ranged around it. The pH of this land would be about 6.0+.
2. Shows the soil starting to degenerate, the hydrogen ions are beginning to displace the calcium, magnesium and potassium. The pH would be falling quite fast and be about 5.0 or lower.
3. Shows the final stage when there is only 1 ion of calcium, magnesium and potassium left — not enough to maintain healthy life. The hydrogen has taken over, and until it can be replaced in its turn by the calcium and the other two minerals, the soil will be unable to grow healthy feed — weeds will proliferate.

ineralization are the only answer to those two as well. *Weeds: Control Without Poisons*, by Charles Walters, is another top book on the subject.

An interesting side-light on analyses similar to the above is that where there is already a good system of livestock management, once the deficiencies are known, the animals do not regress in their health too much. If a basic loose-mix lick like the one given in Chapter 8 is put out for all stock, they will maintain themselves in good health while the land improves. Even when the soil has recovered, the lick should be left out all the year because each year, when new pasture grows after the winter or a long dry spell, there is no real nutrition in the grass for the first six weeks. Dr. George Miller, D.V.M., did much work on this in Gippsland when he was working for the Department of Agriculture.

Soil analysis B — higher pH

The following analysis is from a farm in northern Victoria also having a few problems at the other end of the scale. The difference from the previous is obvious and, as it is the first analysis I have seen from this farm, I would expect that the farmer has been bringing his land into balance for a few years now.

Interpretation

This analysis shows land in good balance needing only gypsum (calcium sulfate) to bring it totally into balance. This will be beneficial on the clay ground and bring the sulfur up to its ideal level.

Depending on the program on this farm, a natural form of nitrogen can be used (compost, etc.) and any form of "biologics" that the owner wished to use. Because the lime minerals are in balance and the pH is ideal, any program using *natural* materials would be both beneficial and economic, the absolute reverse of the previous analysis shown.

Soil Analysis B

Item	Result	Desirable Level
Color: grey brown		
Texture: clay loam		
pH (1:5 Water)	7.0	5.5 - 7.5
pH (1:5 0.01 c12)	6.3	
Electrical Conductivity, EC us/cm	397	<300
Total Soluble Salt, TSS ppm	1311	<990
Available calcium, Ca ppm	2520	1800-2400
Available magnesium, Mg ppm	468	430-600
Available sodium, N ppm	253	<205
Available hydrogen, H ppm	38	< 71
Available nitrogen, N ppm	22.00	25
Available phosphorus, P ppm	44.80	30
Available potassium, K ppm	312	190
Available sulfur, S ppm	3.70	7
Available copper, Cu ppm	5	2
Available zinc, Zn ppm	10.10	7
Available iron, Fe ppm	23	>20
Available manganese, Mn ppm	109	>20
Available cobalt, Co ppm	2.40	>1.0
Available molybdenum, Mo ppm	0.90	>1.0
Available boron, B ppm	1.00	>0.6-1.0
Total Organic Matter, OM %	8.30	6-10
Total phosphorus, TPN ppm	675	
Extractable aluminum, Al ppm	NR	
Cation Exchange Capacity, CEC	17.83	
Exchange Sodium Percentage, ESP	4.95	<5
Calcium/Magnesium Ratio, Ca/Mg	3.23	2-4

Percentages of Ca, Mg, Na, K and percentage of CEC

	Percentage	Percentage of CEC
Exchangeable calcium, Ca	68.47%	56.76%
Exchangeable magnesium, Mg	21.18%	17.55%
Exchangeable sodium, Na	5.95%	4.94%
Exchangeable potassium, K	4.33%	3.59%
Exchangeable hydrogen, H		17.11%

Recommendations:

Gypsum	1 ton to 2.5 acres
Lime	Nil
Dolomite	Nil
Nitrogen	5 kg to 2.5 acres

The term biologics as used by Neal Kinsey refers to the best biologics of all, Biodynamics. It also refers to foliar sprays, seaweed, fish, and some rock minerals — all great aids to continuing fertility.

Soil analysis C

This analysis is included to prove the point that a perfect pH can be a potential disaster area; the farm has too high magnesium and is quite the most dangerous and possibly most difficult to deal with of all three farms shown here. It shows magnesium rather similar to that seen in most parts of the United States — too high in relation to the calcium. It is an uncommon situation in Australia, but it is the reason I have the horrors when people suggest to me that they do not need an analysis and will "just spread some dolomite." If that were done on this farm nothing would work at all.

Calcium cannot be raised too much. The recommendation of gypsum (calcium sulphate) and a small amount of lime should be followed. Later, if there are problems with too much calcium, yellow sulfur will have to be used. This and gypsum are the only ways to raise the sulfur unless a long-term composting program is initiated.

I prefer to use gypsum where there is clay in the soil because gypsum is a great conditioner. The high salt will be reduced by the gypsum top-dressing also, but this can only be done once in this case for the reason stated — the magnesium has to be brought into balance and the calcium kept at a reasonable level. This particular farm is the result of ill-advised and continuous irrigation. The ideal would be to put it back to dryland farming for an extended period until the land is recovered.

Irrigation as a modality will have to be seriously thought out. I have had an irrigation farm and, because the soil was in balance and aerated, I only took three water runs a season, unlike all my neighbors who ran the water through once a week which resulted in leached and damaged land. If farmers stored their own rainwater in dams,

Soil Analysis C

Item	Result	Desirable Level
Color: dark grey brown		
Texture: medium clay		
pH (1:5 Water)	6.60	5.5-7.5
pH (1:5 0.01 c12)	6.10	
Electrical conductivity, EC us/cm	91	<300
Total Soluble Salt, TSS ppm	391	<990
Available calcium, Ca ppm	2720	6158
Available magnesium, Mg ppm	2256	652
Available sodium, N ppm	398	< 52
Available hydrogen, H ppm	103	< 91
Available nitrogen, N ppm	7.10	25
Available phosphorus, P ppm	44.00	35
Available potassium, K ppm	332	270
Available sulfur, S ppm	1.70	7
Available copper, Cu ppm	4.6	2
Available zinc, Zn ppm	2.80	7
Available iron, Fe ppm	38	>20
Available manganese, Mn ppm	11	>20
Available cobalt, Co ppm	2.40	>1.0
Available molybdenum, Mo ppm	0.60	>1.0
Available boron, B ppm	.30	>1.0
Total organic matter, OM %	4.50	10
Total phosphorus, TPN ppm	1372	
Extractable aluminum, Al ppm	NR	
Cation Exchange Capacity, CEC	44.28	
Exchange Sodium Percentage, ESP	4.95	<5
Calcium/magnesium ratio, Ca/Mg	.72	2-4

Percentages of Ca, Mg, Na, K and percentage of CEC

	Percentage	Percentage of CEC
Exchangeable Calcium, Ca	38.88%	30.04%
Exchangeable Magnesium, Mg	53.73%	41.51%
Exchangeable Sodium, Na	4.94%	3.82%
Exchangeable Potassium, K	2.43%	1.87%
Exchangeable Hydrogen, H		22.74%

Recommendations:

Gypsum	12.50 tons to 2.5 acres
Lime	0.60 tons to 2.5 acres
Dolomite	Nil
Nitrogen	18 kg to the hectare

they would be far more conscious of how to use it and do less damage to the environment and their farms.

General comments on this and other analyses

Iodine cannot be shown on a soil analysis at present as it is not strictly a mineral. Farmers should remember that Australia is almost universally deficient in iodine, even Tasmania and all coastal districts. Gippsland is notorious for animals showing clinical signs of an iodine deficiency, but in practically all other areas, stock suffer subclinically as well. This means that they will not be in optimum health. The function of iodine will be covered in Chapter 8. The seaweed portion of the lick at the head of that chapter will provide the necessary iodine in its safest form.

Iodine may be yet another casualty of chemical farming; the United Kingdom, surrounded by sea and quite small, was not low in iodine in the past. It is now since superphosphate and ammonium nitrate have been used for many years.

Priorities

As always, the most important priority is to get the calcium and magnesium in balance as fast as possible. Carbon is the most valuable component of soil; without it nothing works and without the correct calcium/magnesium balance carbon cannot be present.

As explained above, trace minerals are another story. As the health of the paddock improves and the pH starts to equalize, trace minerals often become available to the stock again. It took only 18 months on one of my farms for the cobalt to become available again and I was able to cease supplementation. (This is a very tricky mineral to feed, see Chapter 8.) I was also able to reduce the amount of copper being fed. Clearly, as the health of the soil improves, the missing trace minerals quite often come back into the food chain. In all cases, if cash is no problem, it should be possible to return the land to some sort of health in four to five years, provided that mechanical aids like aeration are also used.

Plant tissue testing

This should only be carried out *after* a soil analysis. Very conflicting results can occur, perhaps leading to wrong remedial measures. Neal Kinsey has pointed out that magnesium levels, for example, can show incorrectly in a tissue test. The plant cannot differentiate between too much and to little; they both show up as a deficiency. Possible remedial action on the soil must not be taken without consulting a soil test. A foliar spray just for the plant could be used in such a case. Tissue tests can vary enormously, according to the time of year, since different minerals are apparently taken up in each season.

Costs

When a farmer says to me that he spent $27,000 ($15,829 US) on superphosphate the previous year and he might just as well spend the same sum on remineralizing, we have no problems. But it is not always that easy. The average cost usually works out at about $140 to $160 ($82 to $94 US) per acre. It never pays to put on half quantities and each area should receive the recommended amount of gypsum/lime and/or dolomite. Some people can afford to do 10 acres, 50 or whatever. But as each part of the farm comes back into balance, the benefits in stock health and fodder quality are so evident that people realize that it is a necessity.

In my farming career, initial remineralization has always been included in the cost of a new farm. It can amount to what is, in effect, a capital outlay since it may be quite a few years before much more needs to be done. Sometimes it is a two-year program if the farm is very sick because, as previously mentioned, there is a limit to what the soil can absorb in a given year. After the initial expense, it is a question of keeping an eye on the levels of the lime minerals and the pH (along with aerating the land). In commercial garden situations, this will apply also with the addition of any side dressings needed. The cost of an analysis varies with companies and I generally pay about $100 ($58.62 US). The important thing is to know what your soil testing service provides in their routine tests so that you can get all the analyses you need without additional tests. It is better to pay more for a reputable company that does thorough testing than to save money but not get what you need which, of course, saves no money in the long term.

Chapter 5

Breeding, Heredity and Environment

Heredity versus environment

A leading cattle judge in New Zealand said to me that it was his considered opinion that the old idea of 50 percent heredity and 50 percent environment (i.e., management) was no longer sound. In other words, it is somewhere nearer 20 to 80 percent environment. It does not matter how good the cattle are genetically, if the management does not match, they will not reach their potential. I have a feeling he is not far wrong.

I can remember my father, who farmed and exported dairy cattle, pigs and sheep from the United Kingdom in the 1930s, telling me that Australia was a fruitful and never-ending source for Britain and other exporting countries. Apparently, after ten years in their new country, extremely well bred and genetically superior animals had degenerated to the point where the farmers had to come back to the country of origin for more beasts to keep up size, production and/or viability. This must have been almost entirely due to ignorance of the prevailing conditions in Australia, where the scarcity of bone-building minerals affected the growth of animals. Geneticists reckon it takes about ten years for a trait, for example small size, to become fixed.

As long as cattle brought into any country are properly fed with all the required minerals and pastured on remineralized farms, their genetic potential will not be lost. It is a totally uneconomic exercise to bring in animals — often at enormous cost — and not look after them properly.

I have had the privilege of talking to owners and seeing some of the Australian Red dairy cattle in my country. They are a splendid type of animal carefully bred up from many European breeds such as Anglers, Swedish, British, and German Ayrshires, Danish and Swedish Reds, English Devons and several more — all obviously carrying the red gene. Shorthorns, British and Australian, and Illawarras have been largely used for upgrading.

Within four generations they are producing a fine, even type of red beast. It is of middle size; its protein production is equal to the best in this country and it is hardy and easy to manage. The breed is a fine example of what informed and scientific breeding can produce. Because of its wide-ranging genetic spread it should not ever, if properly managed, run into problems associated with too pure breeding. I would emphasize the findings of several eminent medicos who tell us that if the correct minerals are missing in the diet, a decrease in viability can show up in as little as one generation (see *Empty Harvest*, by Jensen and Anderson).

As emphasized in Chapter 4, it is absolutely essential that any farm importing stock, either as animals, semen or embryos, see to the mineral levels of their soils. Additionally they must bring the calcium and magnesium in both the paddocks and fodder up to a balanced level as soon as they can. Once the paddocks are fully balanced and the pH reaches a level of around 6.0, it means that every trace mineral on the farm is available to the stock.

The use of phosphorus and artificial nitrogen has to be phased out; both these artificial fertilisers inhibit copper

100 percent (according to André Voisin). The four main bone-building minerals are calcium and magnesium — usually supplied as dolomite as long as the ratios are correct — copper and boron. The latter two are trace minerals and they, with many others, are unobtainable when the soil is out of balance and the pH is too low.

Another result of balanced pH is that undesirable weeds are at a minimum — they prefer a low pH. The other end of the scale is a host of irrigation dairy farms with pHs in the low 4.0s and cattle well below par with, in many cases, a high incidence of Johne's disease (see Chapter 11).

On too many generations of pure breeding

Pure breeding, what is it?

Farmers should realize that pure breeding of farm stock as we know it is relatively new on the livestock scene. Old breeds existed, but they had often been bred out to other breeds if their health or shortage of animals dictated it. When AI (artificial insemination) first became widespread, this problem was highlighted. Having found a way of using prepotent top performance bulls to the maximum, the cattle breeding industry embraced the new technology with enthusiasm.

In Canada it was suddenly realized that almost every Hereford in the country was related to one of nine bulls. Breeding was getting much too close and there was some hasty out-crossing to less glamourous bulls to help correct the imbalance. The reason for the alarm was that lack of viability was showing up in the form of a hereditary tendency to arthritis, previously only known in the breed at very low levels. Even with such a degree of pure breeding, the defect was beginning to be a problem.

Unfortunately, it is nearly always the bad points which seem to come to the fore in unbalanced breeding programs. Nor is indiscriminate crossbreeding always the answer either. It is necessary to take the best of each breed when instituting such a program, as it is no good expecting that the mating of two entirely indifferent animals will produce something good. They can only, in the normal order of things, produce the inadequacies of their parents.

Hybrid vigor is a desirable trait in many cattle programs, especially for beef. I have found that by the third generation it no longer figures to any extent, and possibly these programs should not go any further than that. I was working with a beef farmer in Western Australia, he is an excellent manager and his beasts regularly top the markets. He uses a Gelbvieh/Hereford cross and puts a Brahman cross that he buys from Alice Springs over them — four breeds in effect. His cattle are great types, lean and carrying excellent muscle and the butcher's prices say it all.

The late Bill Howe, past President of the Shorthorn Society in the United Kingdom, Shorthorn breeder and international cattle judge, was adamant that pure breeding could continue too long and viability be lost thereby. In an excellent article in the British Shorthorn magazine, which was published in *Red Cow International*, he explained how various breeds have met the challenge. "We (Shorthorn breeders) pride ourselves in having the oldest herd book in the world. . . . Unfortunately it is this which has contributed to our decline. No breed can continue indefinitely, whether cattle, pigs, sheep or poultry, without an infusion of new blood." He emphasizes performance as the only criteria for breeding animals, and says: "...it is generally acknowledged that the Friesians in Britain have benefitted enormously from their several importations, and from the fact that their black color has enabled the breed to mask much of its mixed ancestry." And a very important statement, ". . . They (new breeders of today) also accept the true meaning of the word pedigree as being a straightforward record of ancestry, it does *not* mean a guarantee of purity. . . ." (italics mine). Occurrences in pure breeds like those which happened to the Canadian Herefords described above are known as "inbreeding depressions."

B. G. Cassell, an American extension scientist, pointed out the results of letting these depressions occur. A one percent increase in inbreeding lowers the milk 50 pounds

Care and feeding of breeders

All the best breeding techniques and bloodlines will be wasted if the heifers and cows being bred are not fed and looked after correctly. Both should be on the healthiest paddocks on the farm coming up to calving. If there is the slightest doubt about the potassium levels in the paddocks, it is absolutely essential to see that they receive extra cider vinegar, at least 100 ml (3.5 ounces) each per day (this should save having to "pull" calves). The vinegar can be sprinkled over supplementary hay or added to concentrate feeds, which animals should be getting by the last six weeks of pregnancy. These feeds do not have to be very large, but they should be top quality in terms of minerals. Pregnancy toxemia-type conditions only occur when cattle are not healthy enough to provide the fetus with all the nutrients (especially minerals) it needs, as well as leaving enough for the mother. The concentrates should contain: 60 grams (4 tablespoons) of the all-purpose stock lick (see Chapter 8) daily. Lactating dairy cows receive up to 80 grams per day in the United Kingdom and Australia.

Cattle should also get a tablespoon per week of cod liver oil, or an A, D and E supplement or high potency injection coming up to calving. This is especially important if the year has been or is very droughty.

A report in *Acres U.S.A.* (July 1994) highlighted the need for sulfur and selenium in pregnant cattle. They reported that at least 1.5 grams of sulfur per day was needed along with 55 mcg per day of selenium. The lick already suggested will cover this, the selenium being sourced from the seaweed meal.

Milking cattle, whether dairy or feeding beef calves, will not milk as well as they should if the precalving feeding has been neglected. Saving money on feed because an animal is not actually producing at the time is extremely poor economy.

per lactation and the milk-fat yield is lowered [
pounds. For example, a bull put over his daughter
percent inbreed and that means 1,250 pounds le:
and 50 pounds less fat per lactation. Even more se
it means up to 50 percent mortality. A bull to his [
ter is a 12.5 percent inbreed and half the above
therefore apply. I heard an Australian, Professor N
from Melbourne University, explain this in 1970 at
nar, but in my particular dairy industry many breec
ply refused to believe it. They learned the hard w

In the beef industry, the enormous size [
esteemed years ago is no longer the desired goa
was a revolution in the ranks about 25 years ago
beef cattle judge at a royal show refused to place
of cattle that had, up until then, been the blue rib
ners. He said, unequivocally, that he was placing t
according to the desired performance on the
other words, the kind of animal that the butc
therefore their customers the general public, we
for and preferred to buy. It caused much heart se
the time, but did indeed make farmers realize th
not necessarily better.

Dairy breeding and replacements for

In the dairy industry, obviously crossbreed
used to its fullest advantage when breeding co\
ing, as the Red Cow fraternity has found. This (
the best bloodlines are kept and viability is
always preferred to line breed back to exceptic
(about four generations back in the pedigrees);
method worked extremely well. If replacem(
required and there is a demand for meat ani
pay to put beef bulls over the cows that are n
echelon. The resulting steers can be sold for i
heifers sold to breed meat animals, their high
tial will ensure they rear strong, quickly growi
these will also be aided by hybrid vigor.

Milking characteristics

As mentioned, cattle can be imported into Australia at present, as long as the very strict quarantine requirements are met. (This is also true in the United States though less necessary). This means that we have a great opportunity to bring in the best that is available from other countries to cross with our local animals. It also means that as we are officially a scrapie-free (bovine spongiform encephalopathy or BSE-free) country, we will also have good export potential.

There are some very fine Friesians being bred here now. Breeders seem to be not concentrating entirely on size as seemed to happen many years ago, but are using the best that North America and Europe have to offer (unfortunately Europe was on hold due to BSE for quite a time).

In the early days, I can remember following a gigantic Friesian bull around in grand parades at various shows. He stood well over me and I am not short; I asked his handler if he was ever used for anything except AI as I considered him too large. "Oh Yes!" was the answer. "He has broken the backs of at least 45 cows." He was used because he had won so many ribbons — something very wrong in any assessment program. Fortunately, the swing from butterfats to solids-not-fat and proteins in the milk has done much to bring farmers around to a saner way of thinking.

Chapter 6

Feed Requirements

Dairy cattle

Cattle dairying has come of age in the last fourteen years. Most farmers have learned that if you want an animal to produce at its peak, then the pastures must be top quality both in type and mineral content. The two do not necessarily go together, but they generally do. Good hay with multiple varieties of herbage in it must also be available when required, and, like the pasture, it must be grown on remineralized land. Crushed barley with, in some cases, ten percent lupins and chaff added makes a good bail feed to which the minerals can be added. If these foods are produced on naturally farmed set-ups, so much the better.

I think that all farmers now realize that dry cows, or those on decreased production coming up to calving, must be adequately fed or they will not be able to produce the huge volume of milk expected without draining their systems of minerals during the early part of the lactation. Unfortunately, in past years there were many who thought that if an animal was not producing milk, it did not need extra feed. Bearing a calf actually makes more demands on the system of a cow than producing milk. If a cow is not fed what virtually amounts to a milker's full amount of minerals during that time, pregnancy toxemia is almost inevitable. The amount of concentrates and hay fed will be dictated by the condition of the cow and environmental factors like the amount of grass available.

Differences in organic and conventional set-ups

In the feeding of dairy cattle there are immense differences according to whether the farmer practices conventional or organic farming methods. In the former, many strange and possibly harmful methods are used to persuade cattle to produce optimum milk quantities. This is often at the cost of their digestion and certainly limits their productive years. Cows fed thus usually have three years active milking life before they are finished.

Those fed on natural, organic principles, while they may not reach the huge production figures of the conventional methods, in the end will give more milk as they will milk on to a ripe old age. Recent work done in the United States suggests that when really well fed on organic farms, cows will equal the outputs of even hormone-fed cows. I certainly found that goats performed at almost twice the rate of those "forcing" their beasts with high-analysis feeds when organics were employed.

In an organic system, overhead will be less and the herd will include cows up to and beyond the age of 10 years happily earning their living. Not much arithmetic is needed, taking into consideration the high costs of replacements, to see which is the soundest method in the long run.

Snags inherent in the conventional system

An article in *Acres U.S.A.* (June, 1993), had a headline that read "Surviving in Dairy." The article, written by Christopher Lufkin, makes the point that the following problems are common to many conventionally run dairy farms.

1. Low water intake.
2. Too high sodium intake.
3. Too high protein intake.
4. Excess non-protein nitrogen.

5. Too much lignin without enough cellulose and hemicellulose in the feeds.
6. pH of the ration is too low.
7. Incorrect calcium to phosphorus balance (and magnesium in Australia).
8. Subclinical anemia is a persistent problem.
9. Too few lactations per cow (see above).
10. Mastitis is another persistent problem.
11. Breeding problems.
13. Not enough sunshine, fresh air and grazing.

Although the author is referring to cattle on the dairy belt in North America, nearly all the points he makes are valid in other countries.

Nutritional needs and associated problems

Water intake

Water should be available at all times and it should not be fluoridated or otherwise treated. Bore water or rain water is suitable, but cattle often prefer the former. These days bore water should be tested as, unfortunately, spray residues and excess nitrogen can end up in underground aquifers.

Sodium

Salt is often added to premixed or pelleted bail feeds, with no reference made to the requirements of the cows. Often licks which contain 80 percent or more sodium are put out with no reference to salt levels in the pasture. Sodium levels that are too high can inhibit the uptake of potassium and other trace minerals. Cattle whose mineral needs are fully met rarely take much salt voluntarily. Salt should be left out in the paddock in the form of natural rock salt for cows to help themselves when they feel they need it.

Too high protein feed

I consider that feeding too high protein would be the main cause of many troubles in conventional dairy farming.

I think 12 percent protein is a reasonable level and rarely produces sick animals. Once it is forced above that percentage troubles start to appear. The incidence of mastitis and acetonemia on some conventional farms is staggering.

I was doing a talk once and remarked that superphosphate was responsible for a great many sick animals. A farmer said that was nonsense, that she put superphosphate on her farm each year and had no problems. As the afternoon wore on, she asked me about practically every illness in the book. When I reminded her of what she had said her answer had me reeling. She said, "Oh, they are not problems; every farmer has those."

Anemia

This is, in Australia, often endemic (see section on iron in Chapter 7). Many soils contain excess iron and are often almost totally deficient in copper. Without the latter, iron cannot be synthesized. So it will be seen that adding iron to the ration is not the answer.

If iron *is* actually too low, it has been suggested by some authorities that adding fish meal or extract supplements a certain amount of iron. There are organically certified brands on the market. Giving cattle a lick, tailored to any deficiencies noted on the soil audit, should fill the bill. The amounts suggested in the lick I recommend (see Chapter 8) can be fed in the bail or be available at lick stations. At *no* time must meat meal be used to correct an iron or protein deficiency (see BSE in Chapter 11).

The lick mentioned in Chapter 8 is a formula that I worked out to be completely balanced. Dairy farmers that I have worked with in England feed 80 grams per day to their milkers. I felt 50 to 60 grams would be enough, but they found more was needed. Some Australian dairies are feeding between 60 and 80 grams as well.

Carbohydrates (bulk)

Cattle depend on having a rumen full of bulky feed; this helps equalize their body temperature. If dairy animals do not receive adequate carbohydrates, which should make up the bulk of the feed, it does not matter how much protein is supplied; they will not keep healthy and supply top quality milk to their optimum. The ideal ratio appears to be about six carbohydrates to one protein. This is not just in the bail feed, but includes the hay eaten during the day. If on irrigation, the grass may not be very high in protein unless a program of land improvement has been in place for a year or two.

Dry-feed levels

In the winter copy of the 1993 *CRT Agri-News*, there was a most helpful table showing the energy value of various feeds. The intention was to point out the best way of creating cost-efficient feeds, but it has a valuable practical application as well when planning a balanced ration.

Energy Value of Feed		
Supplements/Concentrates	B	C
Oats	11.9	90
Barley	13.0	90
Wheat	13.6	90
Maize	13.9	90
Barley Bran	9.2	88
Wheat Bran	10.0	86
Wheat Pollard	11.9	87
Rice Pollard	14.1	90
Peas (23% crude protein)	13.0	90
Lupins (30% crude protein)	13.0	90
Brewers Grain	11.2	25
Hay		
Clover	9.1	83
Good quality	8.5	83
Average quality	8.0	85
Poor quality	7.0	90
Oaten	7.7	90
Alfalfa hay		
Pre-flowering	8.9	85
Flowering	8.1	90

B = MJ (ME) per kg dry matter.
C = Proportion of dry matter (%).

It is of interest to read the energy values of the various feeds. Barley, for example, has the same value as lupins and is far cheaper, nor does it have the unfortunate side effects that can be caused by feeding high amounts of lupins. Barley also has the great advantage of supplying plentiful vitamin B5, needed by all animals to process their own cortisone (in conjunction with vitamin C, a cow makes about 30-plus grams of vitamin C a day).

Rice pollards is an extremely valuable feed and does not have to be fed in large quantities. Care should be taken that it is not grown with chemicals. It is higher in protein than lupins and does not cause iodine depletion as they do. Approximately 200 grams of rice pollards would be enough each day.

Both barley and rice pollards are good feeds. An organic dairy farm I know feeds cracked barley and 10 percent lupins with the necessary minerals (as mentioned below) added. The pastures are also improved and good quality hay is available when necessary. The cows are in excellent order and produce a great deal of high-quality milk.

Legumes

These, whatever kind, should always be fed with care. They are all, even Tagasaste (lucerne) trees, goitrogenic. In other words, they can cause iodine deficiencies. Lupins should not be fed at a rate of more than 10 percent of the given feed at any time. The previous table shows the protein you can expect from lupins these days.

Feeding extra fat

When I am told that the diet also includes used frying oil, I would hesitate to be responsible for the health of a herd fed thus. On the conventional scene there is a vogue for feeding this material (used and recycled) to dairy cattle. Apart from the fact that it is one of the most carcinogenic materials known, and all heated fats come into this category, it does not do anything for the health of the cows. Fat tends to line the gut and inhibit absorption of carbohydrates — most undesirable. Rice pollards would provide extra fat and be beneficial.

Scouring cows

When cattle come in to be milked, they should not have dung all over their back legs. A scouring cow, no matter how great her production, is an animal whose health is below optimum. Over-feeding of protein and pastures whose mineral and nutritional values are not as good as they should be will cause this effect, as will a lack of copper. When analyzed, the pastures often will be found to be completely out of balance and, if conventionally fertilized, this will explain the lack of available copper.

Mineral requirements

If the above advice about feeding the standard home-made lick in Chapter 8 is followed, much of this section is unnecessary. I have quite a lot of dairy clients in Australia currently feeding this lick to their beasts and occasionally we have to juggle the minerals a little until the soils are in balance too. Lack of sufficient sulfur causes buffalo fly attacks in Queensland, so on occasion sulfur needs to be doubled up.

The required amounts of the lick can be added to the premix if desired, then *no* additional minerals should be fed. Often these premixes have large amounts of chelated and unbalanced minerals included which are not desirable.

Extruded feed (pellets) are not as desirable as straight grains and chaffs because it is difficult to tell the quality of the feed that went into them. Also, as they are virtually predigested, the cows' rumen does not have to do enough work when digesting them and, in extreme cases, they lose the ability to chew their cud.

In a country with poor soils like Australia, the dolomite in the lick ensures that mastitis is at an absolute minimum, as well as acetonemia (ketosis). However, from what my consultants over there tell me, this does not work in the

United States as the feed proteins are naturally too high. This is where the copper (in the form of sulfate) in the lick comes into play. Dolomite contains calcium *and* magnesium; the latter mineral is generally in poor supply in all feedstuffs as they are usually grown with artificials, and is also depressed by the practice of feeding high-calcium supplements as well. This can be a dangerous practice. A headline in an English farming paper stated that "Calcium gives cows mastitis" — not strictly correct, but, in effect, it did because it lowered the magnesium in the animals' systems (see Chapter 8 on minerals).

Sulfur

The allowance of yellow powdered sulfur in the lick ensures proper digestion — it will also stop lice should they be a problem. When I was in the United Kingdom, I was horrified to hear that lice were an accepted part of the dairy scene. Farmers have reported to me that the food conversion rate has improved immensely when the rations included dolomite and sulfur as above. Also, as stated in Chapter 8, sulfur is essential for the amino acids in the gut to perform so that the cow can assimilate selenium.

Seaweed

Seaweed, in some form or another, is fed nowadays worldwide. I guess that the quality of feedstuffs everywhere has deteriorated so much that seaweed has become a necessity. It is part of the general purpose lick; it may also be left out for animals to freely take on their own (if financially possible). It is possible to buy high-grade seaweed powder in bulk at a discount for those who use large quantities. If using the liquid seaweed as a drench, make sure to use a source that does *not* add chelated extras. Chelation is undesirable down the throat and I have seen dairy cattle with very damaged livers as a consequence. A tablespoon of seaweed powder or a two ml-dose of liquid seaweed per head per day should be enough. This amount could be halved as the paddocks and the health of the animals improve. Better still is to let the cows help themselves to the lick or powder, which must also be free of added urea.

Additional minerals in the lick

Occasionally, low levels of boron or cobalt on the soil analysis will necessitate adding them to the standard lick in small amounts. Cobalt can be added at a rate of 10 percent of the copper, and boron at a rate of three percent. See chapters on minerals for details on the requirements.

Breeding problems

These are most often due to a defective diet. Paddocks which are lacking in potassium, or where the vitamins A and D are too low (the result of chemical fertilizers and poor paddock health), mean that there will be calving problems. Potassium can be supplemented coming up to calving with cider vinegar in the feed or on hay. Cows that do not cycle properly are low in copper; those that cycle and do not conceive are low in vitamin A due to either drought or defective pastures. Cod liver oil on a regular basis will supply the vitamin. It is important that cod liver oil, and all vitamins, must come in a light-proof container as vitamins are destroyed by light.

Dusty feed

Serious attempts must be made to keep the dust in feed to a minimum or the cows' lungs will suffer, to say nothing of the humans operating the dairy. All grains, if milled too fine, will be dusty (*and* deficient in vitamins). In the United States there have been reports of serious lung damage to cattle from this source. Also, dust is the price we pay for any crop grown on less than really good soil. Years ago dusty grains were returned to the contractor, now it is almost impossible to find feed that is not dusty — another incentive for growing one's feed naturally.

Cows will actually learn quite quickly to digest whole grains. Barley is an excellent source (see section on vitamin B5) and can be mixed with chaff as suggested above. This combination is far less dusty and better for the cow.

Moldy feed

This section applies to all cattle, both dairy and beef. During the past few years my calls about stock affected by molds have been on the increase. Most of these molds have long names, but all have basically the same effect — including death. Any brought-in feed should be examined carefully to see that it is not moldy and to insure that it has not been moldy at any stage. Past mold is easily detected; the feed, if dry, will send up clouds of dark-colored dust when handled. Do *not* feed it to stock; it can be lethal. In damp feed, the vivid and rather unpleasant colors of the mold can be easily seen. This is another inherent disadvantage of pelleted feeds; the molds cannot be seen. Many deaths have resulted from this cause.

Molds are present only on feed that has been grown on very sick land. The great American soil scientist, Dr. William A. Albrecht, explained quite graphically how the mycorrhizae become sick with the fungi, which then attacks the roots of all plants when the ground is sick enough. The fungi then go up the plant and are eaten by whatever stock are unlucky enough to have to graze them — with disastrous effects. Mycotoxins produced by *Fusarium*, aflatoxins produced by *Aspergillus*, sporidesmin, *Pithomyces chartarum* and thiaminase are just a few of the names that have been mentioned. *Hoard's Dairyman*, in an article by Dr. Allenstein some years ago, suggests there are in excess of 200 varieties of mold. But the base cause is always the same — sick land. This is one important argument for restoring the health of the soil. Too many valuable beasts have been lost to these toxins. The damage fungal-polluted feeds cause, if the animal lives, is nearly always hormonal, residual and incurable in the long term.

Beef cattle

Minerals

All the above statements about additives apply to beef as well as dairy enterprises. Whatever the state of the land, they must have the dry mix stock lick described in Chapter 8 available at all times. If it is feasible, the range/fields or whatever should be tested and remineralized as well.

A fully healthy bovine has calcium levels of 2.1 to 2.8 ml per liter blood serum and magnesium at almost exactly half that; the phosphorus should be level with the latter. These tests were done on range cattle in Queensland who were in extremely poor health and no drenches were working. After seven months on the lick already mentioned, they were fully recovered.

The same paper from which the above figures are taken (put out by the Queensland DPI after testing the cattle, which were by then in top health), showed that serum copper (µg/L) should be between 500 and 1100. In that test, the animals which had been on continuous dolomite, sulfur, copper and seaweed supplementation for seven months were nowhere near the top level. It appears that copper deficiencies are becoming almost endemic. This was the first time that the lick was tried and the results speak for themselves.

Magnesium

Magnesium deficiencies can be quite insidious, and will usually appear as tetanies and possible bone deformities. However, the account of the farm where the magnesium had run out should be enough to alert farmers not to ignore this very important mineral. That farm grew excellent looking crops which, because the farmer used superphosphate and lime every year, were totally without any magnesium.

As mentioned above, I prefer to see salt always obtainable for cattle to take as they want. If available, rock salt is ideal.

On big cattle runs, it works very well to put the licks out near the water bores or holes. Otherwise they can be put out in the paddocks. All these minerals, along with adequate feed, will help to keep beef cattle in top health.

Recent research suggests that the natural form of trace minerals, as supplied by seaweed, are far more potent than the chemically manufactured variety. Farmers in range conditions in northern Australia found that the practice of feeding urea to enable cattle to digest very coarse feed was unnecessary once the standard lick was being used.

African-type grasses

Cattle fed on African-type grasses must be given these licks. The majority of these grasses cause calcium/magnesium and iodine deficiencies and horses are very badly affected by them. In horses, the condition is called Buffel head (Big Head, Nutritional Secondary Hyperparathyroidism). The high oxalate levels deplete calcium, magnesium and iodine. This can and does affect males (bulls, rams, etc.) of other species on occasion. It is not worth taking the risk. Below is a table showing which of these grasses are safe (see also section on poisons).

African-type Grasses				
Grass	Total Oxalate	Ca Total	Intake Ca/Oxalate Balance	Estimated True Ca Digestibility
Non-hazardous				
Flinders	.25	1.92	10.0	99
Rhodes	0.45	1.79	46.1	76
Hazardous				
Pangola	0.92	0.17	- 5.7	39
Green Panic/				
Guinuea Grass	0.81	0.32	- 9.7	42
Para	0.75	0.29	-13.4	24
Kikuyu	1.30	0.23	-22.4	20
Buffel	1.42	0.22	-22.6	16
Narok Setaria	1.81	0.15	-14.2	32
Kazungula Setaria	2.82	0.97	-30.3	3
Ca = Calcium				

Urea

Except when it is a matter of life and death, as when feeding quite indigestible feeds in drought conditions, avoid using urea. It is a poison and the signs of urea poisoning and antidote are listed in the section on poisons. It is also inadmissable in feed if seeking organic certification. Given the information above, using the standard stock lick means that feeding urea is unnecessary anyway as the ingredients in the lick enable the cattle to make full use of the forage.

Feeding minerals with hay

For those using the big roller bales, which most farmers do these days, feeding the necessary minerals is very easy. A farmer I know in New South Wales tried this at first in a bad drought and reported great success. When the bale has been unrolled, he drove along it with his spreader containing the lick mixture. Cider vinegar, suitably diluted (ratio about 1:4) can also be added by spray for cows coming up to calving. This method ensures that all the animals get what they need. This farmer found that he only needed to give his cattle access to these extras about every ten days — not each time he fed the hay.

Chapter 7

House Cows
and Rearing Calves

Acquiring the first cows

Unless you have experience in cattle farming and handling, the general market is possibly the worst place to buy your initial stock. I did just that, buying the crosses I wanted, mainly Shorthorn over dairy, regardless of the condition in which they were presented. It was not very good. But to me it was an exercise to show myself that if the genetics were there, my management challenge was to make the best possible use of them. It was an uphill battle at first, but the end product was worth it. They were a slightly disparate lot in appearance, but all did extremely well after rather doubtful starts. This is not a job for one new to farming.

All these animals were bought with the object of selling them for beef. I sold them at about two years; the last 14 months of that time they received absolutely nothing except the improved pastures of the farm. Their sale prices fully justified the exercise. If you are thinking of a milking cow or starting to build up a small stud operation, make sure that the source of your cattle is a farm free of Johne's disease (see Chapter 11), and that the cattle look well and happy. It is no good buying trouble.

Priorities

First establish your priorities. Do you only want one house milker or are you regarding your purchase as the nucleus of a beef or milking enterprise of some kind? With that idea in view, do some homework on the different breeds, their characteristics, advantages and disadvantages. When you have worked that out, either go to the organization or society of the breed in which you are interested, ask an agent for help, or, if you know a farmer already involved in that particular breed, enlist his help.

Prices

The price of the stock will be in direct relation to its breeding and background. Calves from the market or "over the scales" will be rock-bottom prices; the latter are usually the cheapest. If you are aiming at pedigree stock, you may well pay four to ten times that amount or more depending on how fashionable the breed happens to be. Regard the purchase as an investment and make sure that it is sound and that you look after it correctly. It is wise, if buying good pedigree stock, to enlist the services of a sympathetic vet and have the animals vetted and blood tested before purchase as well.

Naturally a well bred cow will cost a little more, but against that her calves will also be worth more. It is actually a fairly cheap way to build up a small herd of good milkers or beef producers. Running the herd on organic lines and if possible, getting certified with one of the certifying groups (there are three in Australia) is also worth considering. There is always a demand for organic meat, and people are often prepared to pay a premium for it.

House milkers

It will pay the farmer to look around for a cow from a breed that is known for quiet handling, something like a Brown Swiss or a good Red Breed cow. Jerseys are great milkers and, when quiet, fine. But they have a reputation, not wholly undeserved, for being intractable. The cow may be quiet enough, but breaking in her calf can sometimes present problems.

Handling

A cow should be halter trained if bought as a house milker. Teaching an adult animal to lead is not a job for an amateur. Make sure that any calf you rear for that purpose is taught to lead when young. The best house milkers are usually calves that have been hand reared, and have, therefore, been taught to respect and like humans. Too often house cows that are broken in late tend to be unpredictable and not very safe for children or very slight people to manage single handedly — the very people who may be looking after them in many cases. There are still plenty of families who live in the country and have a house cow. I have a house goat, preferring my milk to be as good as possible. Hopefully these days the number of home milkers may even be growing.

Obviously it is better to buy a house cow that has been broken in if you are not experienced with cattle. Otherwise you will need a good, strong bail as shown here and probably some help as well for the initial efforts. Once a cow learns that she is to be fed in the bail, the battle is half won. When really quiet, they can be tethered and milked in the open quite easily.

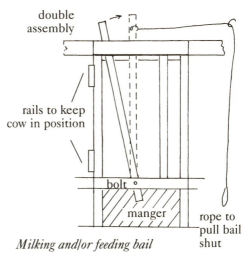

double assembly

rails to keep cow in position

bolt

manger

rope to pull bail shut

Milking and/or feeding bail

It is a good idea even with one house cow, unless it is trained to walk into a float, to have a small yard and race for loading and handling. The race needs to be just wide enough for a cow to stand in so that she cannot turn round. It should be possible to restrain her should she need attention and also so she can be inseminated in it if necessary. It is also a good idea, if possible, to have a portion of the yard under cover, especially the handling races. It is quite often necessary to treat or handle cattle when it is pouring down rain and there is no sense in making it more uncomfortable than it needs to be for either cow or human.

A simple handling system is show below.

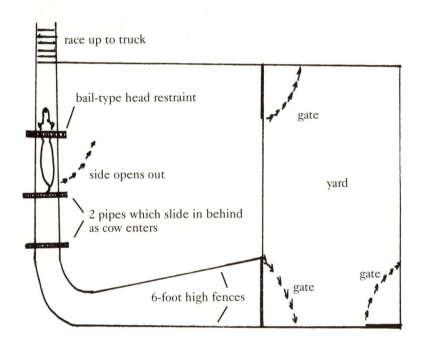

race up to truck

bail-type head restraint

side opens out

2 pipes which slide in behind as cow enters

6-foot high fences

gate

yard

gate

gate

Rations: an easy bail mix

A good homemade bail ration for a house cow is made up of the following base ingredients:

    one part alfalfa chaff,

    one part oaten chaff,

    one part bran,

    and one part rolled barley or soaked whole barley.

To this mixture should be added daily: a tablespoon of the lick described in Chapter 8, but the cow should also be able to help herself to more of the lick in her shelter if she needs it and cider vinegar, roughly 50 ml per day (almost two ounces), should be added to the water soaking the barley.

A week's supply of the above mixture can be made at one time. A two-gallon bucket of this ration can be fed once a day at milking. If the cow is feeding a calf, the owner will only milk her in the morning and the calf will milk her at night, or *vice versa*. When the calf is not suckling, it has to be tethered so it cannot reach the cow, which may mean tethering the mother as well. If the cow is to be milked twice a day, split the ration in two. Good pasture hay should be available if the pasture is poor, otherwise the cow should do well on the above ration plus grazing.

Look in the section on copper for indicators that your cows may need to be supplemented. Many red cattle look quite orange when their copper levels are too low, black ones tend to look rusty, and all cattle have fluffy rough coats, often with a little curl at the end, when they need extra copper. However, the amount in the lick should have her appearance improving very fast.

Fostering calves

This is another alternative to milking the cow yourself; your cow may be used as a calf-rearing machine. Some cows take to nursing strange calves quite easily, others need more persuading. It is a good idea to fasten the new calf to the cow's own offspring. A stout leather collar on each calf, joined by a short strap or rope, will accustom the cow to the smell of the new calf fairly quickly. Once they are both sucking alright, the foster calf can be let lose. A big, old Friesian cow belonging to a friend had, at last count, reared 30 calves — she was a wonderful mother.

Joining

Before the time comes for the cow to be bred again, make up your mind which kind of bull you wish to use. Do you wish to rear another house milker for sale or your own use, or is the requirement for a beef calf which can be put in the freezer or sold to augment the farm income? Having decided which plan to follow, have a look round. There may be a neighboring farmer with the type of bull in which you are interested. Find out if your cow may join his mob for three weeks and what kind of service fee will be needed. If that is not practical, consult with a local artificial insemination (AI) service so your cow can be inseminated at home. By choosing AI, you have the choice of a wide range of bulls, but if you are starting with an empty heifer, do not choose too large a breed to put over her. She should be rising 18 months at least before being mated, unless she is very well-reared and grown.

Having had your cow mated, continue to milk her until she dries up. Depending on her breed, she may milk through until she calves again. Make sure that she is fed her proper ration during that time because from the sixth month of pregnancy she will be doing double work. It is important to see that cider vinegar is added to her diet for the last two months before she calves, unless it is already part of the daily ration. She should also be on a tablespoon of cod liver oil once a week mixed into her feed. If she is dry, accustom her to being brought in and fed in the bail from about the seventh month (or before), so that she will be used to being handled when she does finally calve.

As mentioned in Chapter 5, the pregnant heifer or cow should be fed, or have obtainable, the mineral lick as well as a feed of chaff and bran (see page 82), coming up to calving. It is the best insurance there is against the ills that can arise around calving time. Consult Chapter 8 on minerals and the lick and Chapter 11 on ailments and it will become obvious that illnesses generally are not inherent, merely the result of poor management.

When she is due to calve, leave her alone, merely checking that she is alright at intervals, and feed her as normal. Remember that sometimes even the most placid cows become very jealous mothers when they calve. Do not take unnecessary risks, particularly if she has horns.

If the cow does not seem to be calving as she should, yard her and tie her up so that you can investigate what is happening. You should have the crush shown below. Insert a *clean* hand into the cervix and see what portion of the calf is in engagement — hopefully two front legs. Feel carefully, if you have never done this before. Strong men have had their arms broken by the contractions when their arm was inside a cow. I once spent five hours hand pulling a dead calf for a neighbour, I could not use my arm for a week it was so stiff and sore.

If it is obvious that you are going to have to pull the calf, the diagram below shows a very simple device much used by farmers here in the early days (and now), known as a Spanish windlass. With it the calf can be quite slowly winched out, it does not impose the strain that using a tractor or a mechanical device often does. If it is obvious the calf needs turning (although this is very unlikely if the cow has been properly fed as above) call in help the first time, a vet or an experienced cattle farmer.

Once the calf is born it will need to be fed immediately, either from its mother or from a bucket into which the colostrum from its mother has been milked. Watch the cow

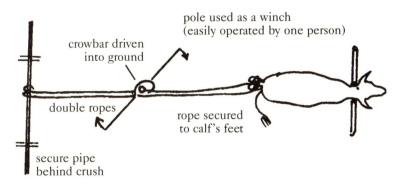

pole used as a winch
(easily operated by one person)

crowbar driven
into ground

double ropes

rope secured
to calf's feet

secure pipe
behind crush

Spanish Windlass (seen from above)

when releasing her from the ropes; it is wiser generally to do this from the far side of the fence. She can be as mad as fire after a calf-pulling exercise on occasion.

After calving, continue the normal routine if possible, getting her into the bail to feed, and bringing the calf up beside her. From this stage it will pay to start handling the calf quite firmly. One advantage of leaving the calf on the cow for one feed a day, is that if the owner has to go away, the calf can milk the cow quite satisfactorily. A lactating cow must be milked twice a day initially at least. Teach it to lead and go where you want it to go. The temperament of its father may make a difference here, and I would recommend that a docile breed is used.

Rearing calves

If you wish to have all the milk your cow produces for home consumption, it will be necessary to hand rear the calf and it is probably wiser to remove it from the cow once it has had the initial colostrum. Colostrum could, of course, be milked out by hand and fed. It is quite essential that the calf has all the available colostrum since it is the best way to guarantee a healthy immune system. Colostrum from another cow will *not* do; the mother confers a unique immunity to her own calf for the conditions in which they live. It is also an unwise practice to feed a calf from another cow, if avoidable. Artificial colostrum is quite easily made up. Two pints of warm, clean milk from a healthy cow, a dessertspoon of cod liver oil and a tablespoon of liquid seaweed (urea free and containing *no* chelated minerals), mix this well, and give in a bottle or teach it to drink from a bucket straight away. It will need about a quart of milk the first day and twice a day thereafter. The amount can be increased as it gets bigger. If buying milk, make sure it has no tallow added.

Feeding poddy calves

Calves need about two liters of milk a feed, minimum; they can be given twice as much, but will grow very well on two liters twice a day. Underfeeding is always better than overfeeding provided the animal is not starved. The ration can be cow's milk, goat's milk, or milk replacers. Calves do very well on goat's milk, although it must be from a CAE-free goat (caprine arthritis-encephalitis is an auto-immune disease of goats which is spread by milk). I prefer to feed calves in a bucket or trough. The initial teaching takes a day or two, but the ensuing feeding processes are much quicker and easier.

To teach a calf to drink from a bucket, hold the bucket up to its head and put the left hand fingers into the top of the mouth so that the calf sucks them. Do not let your fingers reach the back biting teeth (cattle do not have top front teeth). Lower the whole hand with the calf sucking into the milk, and keep on doing this until the calf drinks on its own. Usually two or three sessions are enough. If converting bottle-fed calves to the bucket the same process will work, just hold the teat in the milk.

Calves and their minerals

It is far easier to give minerals to bucket-fed calves. The minerals can be put in the milk and the calves will lick them up as they always go on licking the bottom of the container after they have finished their milk. A dessert-spoon of dolomite per calf twice a week, and two ml of sea-weed concentrate on the same days, is usually more than enough. A teaspoon of cod liver oil in the milk once a week is also a good idea. The calves often do not lick up all the dolomite straight away; it can take a day or two. Feeding minerals means that calves should not get calf scours, which are due primarily to a lack of magnesium. The sea-weed gives them the trace minerals they need.

Calves and concentrates

Once they reach the stage of having some concentrates, the minerals can be fed with them. When I reared poddy calves they went straight into the paddock once they were drinking well. They were called to the fence to drink. It is necessary to have a chain and clip at each calf's drinking station, otherwise the fast drinkers take the lot. The milk can be stopped at about two months, although more will not hurt. I did not buy any made-up calf feed, merely gave them hay when needed and a mixture of lucerne (alfalfa) chaff, oaten chaff and bran with their minerals added each morning. This also makes them easier to manage — they'll follow anyone with a bucket. When they reached about six-weeks old I gave three of them a five-liter bucket of this mix once a day between them. This was kept up during the winter. They were reared on an organically converted farm and received no supplementary feed after they were seven-months old. They grew on to be very fine-looking animals, ahead of their age in size, in spite of being rather poor when I bought them.

Managing calves

Make sure the calves are taught to respect fences when young. If they have been able to scramble through fences, they never forget it and will continue doing it all their lives which is totally impractical. One cannot really stop cattle jumping fences, all of them will do it if there is enough incentive — like a bull on the other side of the fence or other cattle if they do not have enough company. It is no good keeping a bull unless he has enough cows to keep him busy; ten is somewhere near the minimum number. If he does not have enough company and/or females, he too will go and look for them. The best strained fence (or electric wire) in the world will not stop him.

Chapter 8

Minerals and Their Uses

More and more authorities are now emphasizing the great importance of trace minerals in the cattle diet. In feedlots in the United States, chelated minerals are the latest aids. It is to be hoped that the cattle arc not consuming them long enough for the liver damaging effects to show up as they did in a dairy farm discussed below.

Experts are pointing out that deficiencies in minerals are one main cause of problems in the cattle industry. However, since writing the first version of this book ten years ago, I had an experience which taught me the dangers of chelation. Chelation by mouth is neither safe or effective. In 1996, I was trying to help a 300-head Friesian/Holstein milking herd in Southwest England. The cows gave every appearance of trying to die on their feet and the farmer was a very worried man.

There was no way of getting a comprehensive soil analysis in the United Kingdom at that time. I had established, by the health of the cattle, that both dolomite and gypsum were needed since the animals had milk fever and lice; but I felt something else was gravely amiss. The farmer's vet took 10 random blood tests a month, so I asked to see the last six months; they literally looked as though they had been rubber stamped — calcium and phosphorus too low — nothing else.

In actual fact, the calcium was slightly above the desired level in the blood; the phosphorus, which should have been about half the calcium, was far too high; the magnesium, which should have been equal to the phosphorus, was way too low (see chart to follow). But what drew my eyes were the bilirubin tests on all of the samples. I asked what the vet had said about the fact that they were 0.5 and below, instead of the desired 7-10. The answer was nothing. One thousand pounds sterling ($1,562 US) per month was spent on highly processed feed and nearly all chelated minerals went through the feed. I suggested that he change to the lick suggested in this book and his reaction was: "That will be incredibly cheap, will it work?" I said it could not be worse than it was at present; the cattle were on their way out.

I had to return to Australia before the farmer could implement the change and I got a letter a week after I got back saying that perhaps he was imagining it, but the cows had been on the lick minerals four days and they looked better. End of story — they went from strength to strength and he is the farmer that now gives the cattle 80 grams of the lick per day.

One other change had to be made, the lick had to be made available to the springing cows. Their calves were becoming totally fly-blown within 48 hours of birth. Since then I have asked many of my farmer vet clients their opinion of chelated minerals and the first reaction was, "Quite dangerous for man and beast, and don't you forget it." Several doctors have also given up prescribing them. In the words of a friend who has been farming and is doing a degree at Aberystwyth Agricultural College in the United Kingdom, "They are to the body what N-P-K fertilizers are to the soil; they work too quickly." Exactly. Chelation in the blood seems to be fine, but not down the throat as it appears to go too fast to the liver.

Another disturbing finding in several recent research papers is that bullets for various mineral deficiencies are not always entirely efficacious. In smaller ruminants, both bullet and scratcher have been found to be coated with calcium, thereby making the nutrients in the bullet unobtainable. Cobalt bullets have been mentioned especially. If you are using this method, make sure, by taking blood counts if necessary, that they are working.

Another authority points out the uselessness of proprietary salt licks and blocks as they never contain enough of the minerals that are really needed. On-farm supplementation, according to soil analyses, is suggested as the most economical method for correcting problems. That is the only way to be certain that the supplementary minerals are tailored for the farm in question, according to cattle health and the soil analysis specific to the individual farm.

The following figures are from a print-out on some Droughtmasters that had been supplemented for seven months with the lick described below. The exercise was monitored by the Queensland Department of Primary Industries. They tested the 10 best-looking cattle. The licks were based on dolomite and contained copper, sulfur and seaweed.

Mineral	Units	Levels	
Calcium	mM	2.5 -	2.8
Magnesium	mM	0.65 -	1.30
Phosporus	mM	1.13 -	2.25
Copper	µg/L	500 -	1100
Zinc	µg/L	500 -	1200
Total Bilirubin*	µM	1 -	10

*Bilirubin is an indication of liver function.

These are the main indicators for general health. The calcium to magnesium ratio is of interest when one bears in mind the extremely uneven amounts of both minerals generally found on any farm. One maxim should be borne in mind; the saying, "If a little is good, more must be better" does *not* apply to minerals.

Pat Coleby's Base Lick Recipe

The base lick, for use either on demand or in bail feeding, is:

50 pounds of dolomite
8 pounds of copper sulfate
8 pounds of 90% sulfur dust
8 pounds of urea-free seaweed meal

The addition of other trace minerals would depend on the soil analysis. Zinc, boron or cobalt might have to be added if their absence was causing trouble. Seaweed contains all three, as well as selenium, iodine and a host of other minerals and is often all that is needed. It is best to start with this basic lick and watch how the cattle do.

This lick should be well mixed (a cement mixer is useful here) and put out in a container protected from heavy rain. Two or three inches of rain on this lick can mean that the copper will be lost in half an hour. Tests have shown that it is in some way inactivated by the dolomite when it all gets too wet. An ounce of prevention, in this case, is worth a pound of cure.

Boron

This is an important trace mineral needed in very small amounts. Calcium and magnesium will not be correctly utilized if it is missing, leading to problems associated with deficiencies of these minerals, like arthritis. Boron is extremely important in the cultivation of legumes. Alfalfa will not grow or create nodules in a boron (or calcium and magnesium) deficient paddock, and root crops will go black inside. Boron-deficient wheat does not fill at the bottom of the ear.

In areas that are boron deficient, like the land around Bendigo in Victoria, arthritis in all stock is widespread. This is usually controlled by feeding either seaweed products, which contain natural boron, or if that is not enough, adding borax. This was the case in my milking goat herd; I had to add a teaspoon a week of borax per 30 goats. Until this was done, the creaking joints were audible as they walked around the paddocks. For dairy cattle, seaweed (either in the bail feed or in the lick) should be adequate as they do not have the high mineral requirements of goats. Even so, we have found herds where boron had to be added once a week at a rate of about five grams (a teaspoon) per head.

Calcium

Calcium is required for the nervous and muscular systems, normal heart function and blood coagulation. It is also needed for bone growth. Calcium *always* must be considered in conjunction with magnesium; the two minerals interact and must be kept in balance at all times. An excess of calcium will cause magnesium to be depleted and *vice versa*. It is, therefore, unwise to feed calcium carbonate (ground limestone) or dicalcium phosphate (DCP) on its own as this would cause a depletion of magnesium. Where magnesium is very high in the soil, the blood should be monitored to check if the levels are correct.

If hand feeding, dolomite, which contains both calcium and magnesium, should still be used. This is because feed is frequently grown either with superphosphate, which ties up magnesium, or on magnesium-deficient ground. (Magnesium was always plentiful in the United Kingdom before the advent of chemical farming). The previously mentioned headline in an English farming paper read, "Excess calcium gives cows mastitis." This statement is not strictly correct. Calcium does not cause mastitis, but, by depleting magnesium, which is needed to maintain udder health, an imbalance was created. Mastitis organisms were then able to gain entry and proliferate.

Excess calcium, both in the plant and animal world, is linked with a weakening of the cell structure and lowering of immunity to disease, especially of viral origin. Dr. Neville Suttle of the Moredun Research Institute, Edinburgh, also emphasizes "that feeding an excess of minerals like calcium can do more harm than good, predisposing to milk fever." Quite.

In Australia, a cow fed the correct amounts of calcium *and* magnesium is as mastitis resistant as is possible. However, in countries where the soil is of better quality (which means virtually *all* other countries) copper must be added to the ration as well to prevent both mastitis and ketosis-type illnesses. Hence use of the lick provides a broad spectrum of safe and natural minerals.

Calcium should be found in all feeds, alfalfa in particular if well grown (i.e., without artificials, irrigation and with the correct minerals). However, its presence depends on two factors: 1) that the original soil where the feed was grown contained adequate levels of the mineral, and 2) whether artificial fertilizers were used; these reduce the levels of available minerals in the feed. For instance, irrigation-grown alfalfa is usually expected to have several cuttings in a year and it usually needs artificial help. That always means that the roots will not go down the usual 14 or so feet that dryland alfalfa does, hence the shortage in minerals.

Conditions caused by lack of calcium are arthritis, uneven bone growth, knock knees, cow hocks, poor teeth, a general lack of well being and susceptibility to cold. Calcium and magnesium deficiencies will also cause lactation problems such as milk fever, mastitis and low milk production.

Occasionally, in cases where there is an excess of magnesium in the soil, calcium may have to be fed on its own. This does occur in some districts (around Moama, Victoria, for example) and shows the importance of having a soil analysis of your farm. Dicalcium phosphate (DCP) should only be fed in the *very* short term. For permanent feeding, ground limestone (calcium carbonate) is safer. In the United Kingdom, where magnesium is (was) also in good supply, the basic lick did not provide too much magnesium because superphosphate and ammonium nitrate had depleted that mineral.

It is important to remember that calcium (and magnesium) assimilation depends on adequate boron, copper and vitamins A and D in the diet. Boron is found in seaweed products; if it is missing on the land (which is quite frequent), it can be added to the previous lick at a rate of about half a pound. Vitamins A and D should be available from the sunlight and well-grown feed, otherwise cod liver oil or an A and D liquid will have to be given as a drench or injection.

Magnesium

In modern conventional farming, magnesium deficiency would be the biggest problem in Australia. Magnesium appears to be more readily inhibited than calcium by artificial fertilizers. Early experiments indicated that the use of two bags of superphosphate inhibited the uptake of magnesium by five pounds of magnesium to the acre. There is also the possibility that many districts were originally deficient in the mineral as well. There is usually considerable leaching of both calcium and magnesium (and other minerals) in high rainfall areas. Also, according to Dr. Harold Willis (a researcher who writes for *Acres U.S.A.*), magnesium is lost both in the body and in the surface of the soil when the temperatures go above 100°F. I wrote to ask him about this as I had found the phenomena very common here.

Between 50 and 70 percent of ingested magnesium is needed for bone growth. The balance is required for neuromuscular transmission, muscular health and a healthy nervous system. The enzyme systems of all animals depend on an adequate supply of magnesium to work properly, which explains why so many dairy cattle farmers have told me that they felt dolomite had improved their feed conversion rate, as well as making mastitis and acetonemia a thing of the past.

Both magnesium and calcium are rendered inert in the body by sodium fluoride and so they become unavailable to the animal, according to information from the United States. This must be considered if animals are drinking water containing fluoride salts, as magnesium is needed for the enzymes to function correctly. It is safer to use rain or bore water without fluoride or chlorine.

As previously mentioned, magnesium can be depressed by an excess of calcium. It is also almost totally removed from the system by feeds high in nitrates, such as capeweed, variegated thistle and some broad-leaved plants, even clover on occasion. Feeds high in oxalates, like African grasses, have a similar effect (see Chapter 6). Cattle grazing capeweed (*Arctotheca calendula*) must have dolomite or magnesite added to their ration. Cattle, in particular, will succumb to scouring and general debility in the first instance and iodine deficiency in the second. In either case, death is often the next step.

Conditions induced by a deficiency of magnesium (with calcium) include grass, lactation and travel tetanies, mastitis, acetonemia, arthritis, founder, warts, soft teeth, bent/deformed bones and nervous behavior, as well as most of the conditions related to calcium deficiency, in particular, uneven bone growth. Many of these complaints have a causative organism; however, it only operates when the host is magnesium deficient.

Animals whose bones have shown abnormal growth patterns or changes have been much improved in a few months by supplementation of dolomite or magnesite. Should a calf that is born in good order later develop a deformity, the cause can be a magnesium/calcium imbalance. Deformed jaws and crooked legs are often due to this imbalance, even at birth, but they are not congenital as often as we believed. Horse's foals have been born with splints, a calcium/magnesium deficiency condition, when mares were inadequately fed in the final stages of pregnancy. These conditions were reversed by properly supplemented feeding.

Animals showing excessively nervous behavior have become much easier to manage when dolomite or magnesite has been added to their rations. They have changed from excitable to quite calm individuals in a matter of days or even hours.

It should be remembered that overdosing with calcium and magnesium can lead to depletion of copper, cobalt, zinc and most other trace minerals except possibly iron. However, if oral copper poisoning is suspected, dolomite and vitamin C and injections of vitamin B15 are a very effective antidote.

Magnesium normally should be obtained from feed grown in soils containing adequate levels of the mineral, which should happen in the United States unless the feed is grown using artificial fertilizers or on depleted land. In the United States, magnesium is not often naturally deficient, nor was it in United Kingdom. Blood tests are the best yardstick by which to measure levels. Healthy blood levels can be found in the chart found earlier in this chapter.

In Australia, there is another source of calcium and magnesium for those lucky enough to live in the right areas — the artesian bore. Analyzing the water would be the best method of finding out what minerals a bore contains. They can vary immensely even in the same district and can also, sadly, be contaminated with spray residues and nitrates.

Dolomitic lime, if of decent quality (around 50 percent calcium and 25 percent magnesium), is the best way of supplementing these minerals, whether in the feed, licks or as a top-dressing.

Cobalt

Cobalt is needed for healthy bone development and for the health of red blood cells. Cobalt anemia can cause persistent ill-thrift, depleted appetite and susceptibility to cold. A subnormal temperature and any or all of the above conditions can indicate a lack of cobalt. Lack of bone growth in the young can be caused by cobalt deficiency and, if not rectified, can lead to wasting and death. Information in the December, 1992 issue of *Acres U.S.A.* shows that a lack of cobalt (also manganese, copper and iodine) allow *Brucella abortus* (Brucellosis, Bang's disease) infections to proliferate. This highlights the importance of having the soil analyzed.

Causes of cobalt deficiency are overuse of artificial fertilizers, use of antibiotics or, very occasionally, an original deficiency. Cobalt is synthesized into vitamin B12 in the gut (like iron). Adele Davis, the great nutritionist, explained this was controlled by the intrinsic factor. This synthesis ceases to work in cases of extra stress as well as for the reasons mentioned above. A cobalt collapse ensues and the beast will have a subnormal temperature and go down very quickly. Lethargy and loose manure are often the first signs. When this happens, the quickest and only way to reactivate the synthesis is to inject vitamin B12. This is a water-soluble injection which is completely safe; the body merely excretes any that is administered in excess of requirements. A large cow needs about 20 cc intramuscularly and a calf needs 10 cc. Vitamin B12 is a great help in encouraging all ages to eat after illness or stress, and should, in fact, be given as a routine measure on these occasions (see section on vitamin B12).

Cobalt sulphate is extremely toxic and should not be given at a rate exceeding a quarter of a teaspoon (1 gram) per adult cow per week unless prescribed by a vet. However, cattle appear to utilize cobalt bullets, although opinions differ on that, too, nowadays, while goats and possibly deer do not. The vet will advise on this. Seaweed products can be fed in the ration or *ad lib* and, as they contain natural cobalt, this will often be enough to correct any problems.

If the land is cobalt deficient, the best course is to have this amended by top-dressing. However, consult Chapter 4 on soil deficiencies, because pH levels must be raised first. Top-dressing an acid soil with cobalt, or anything else except lime, dolomite or gypsum, will be a waste of time. Often when the pH is raised by these means, the cobalt becomes available again.

Excess molybdenum will inhibit copper and cobalt. It is most important to make sure any analyst who tests the farm's soil includes cobalt in their test. Copper deficiencies are serious, but appear to be much worse if cobalt is also missing. It is important that a farmer knows the mineral status of the fields on which the cattle graze.

Copper

Copper is needed for optimum health, resistance to diseases — especially those of fungal origin — and protection against internal parasites. Keratin, the basis of hair, skin and hooves needs copper (and sulfur). Failure to come into estrus regularly is possibly the most damaging result of copper deficiency from a financial point of view. Cattle that receive the right amount of copper cycle regularly and at the correct time.

Foot rot and associated foot problems are serious conditions caused by too low copper. Veterinary manuals going back to 1919 cite copper, used both topically and in the food supply, as the necessary mineral to combat foot rot. This is still the case. Foot rot and its precursor, foot scald, can be brought under control by making sure the stock have adequate copper supplementation.

Copper's most important role in the body is to assist in the absorption of iron. This *cannot* take place without copper. Australia is almost universally high in iron, yet anemia is the most common condition in every type of livestock. The enormous range of iron tonics available in all feed stores testifies to this fact. Ironically, nearly all the analyses that show high iron are accompanied by almost equally low copper in each case. It is quite useless, and dangerous, to administer iron tonics to rectify anemia without treating the basic cause, which is often a lack of copper or the result of severe worm infestation.

Signs of copper deficiency in all animals are therefore: anemia, scouring, heavy worm infestations, and lack of color in the coat (black-colored cattle turn a rusty red; red ones turn orange). Fungal conditions such as cowpox, foot rot, and ringworm are caused by low copper and these are often linked to herpes-type outbreaks which, like staph infections, only occur in copper-starved beasts. Cancer and low natural immunity are also due to a lack of the mineral, as found by André Voisin.

Black and dark-colored cattle develop staring coats with a little curl at the end of the hair. Many of these conditions can also be occasioned by bad worm infestations as well. A friend with a fine Friesian milker destined for the state show noticed the coat looking rough so dosed it with the most powerful drench on the market. Six weeks later, after restoring it to health, he noticed the curl at the end of the hair and gave it some of the above lick and it was right in a few hours.

In *Hungerford's Diseases of Livestock*, the author says that repeated scouring is often caused by copper deficiency. Johne's disease and Brucellosis do not occur in animals whose copper levels are correct.

Occasionally cattle and other stock will show the classic "spectacles" appearance when copper deficient. The skin around the eyes appears light colored and pulled away, making the animal look as though it is wearing spectacles.

Lack of copper will also cause cattle (and all stock) to chew fences, strip bark off trees, eat metal objects and anything else available in their efforts to find enough of the mineral. Copper deficiency is the cause of the infamous "hardware disease." This is the term that was used in Australia for cattle that kill themselves by eating strange metal objects.

Copper deficiencies are largely caused by application of too much chemical fertilizer. Superphosphate and nitrogenous fertilizers both have this effect (according to research conducted by André Voisin). Molybdenum renders copper unobtainable and, according to Dr. Neville Suttle, the molybdenum suppression of copper affects cattle more than sheep. In some districts, molybdenum fallout from industrial complexes appears to be increasing and the result is a lack of copper. As mentioned in the early chapters, an excessively low pH as well as leaching can be the cause of many trace mineral deficiencies. Large amounts of zinc also inhibit copper, but the reverse does not happen.

Copper is highly toxic, but the requirement seems to be much higher than has previously been considered. Cases of copper poisoning should not occur if supplementation is carefully monitored and if the copper is always supplied with dolomite (as in the lick mentioned previously) which appears to prevent toxicity. Too much copper kills, but too little does just the same. For this reason, copper injections are not a good idea because an overdose cannot be reversed.

Copper sulfate, not carbonate, should be used for cattle supplementation as it appears to be more effective and safer. Copper carbonate is not so readily dissipated and an excess could be fatal; it is best used as a top-dressing on the land. John Goode, a producer from Kingston, South Australia, who spoke at the 1993 J. S. Davis Beef Research Forum stated "Although copper can be lethal in large doses, we have not had any deaths lately despite using quite high levels." Seaweed meal contains a significant amount of copper, and very often adding the meal to the ration is enough to make up a small shortfall of the trace mineral.

In severe cases of copper poisoning, give dolomite and an equal quantity of vitamin C powder (sodium ascorbate) in the mouth every two hours, and administer injections of vitamin C and vitamin B15 in the same syringe every few hours until recovery. For a calf use 10 cc of vitamin C and five cc each of vitamins B12 and B15; use three times that amount for a large beast. In a bad case of copper poisoning, a vet found that adding activated charcoal also helped.

Iodine

The whole of Australia appears to be iodine deficient, surprisingly even in coastal areas. This is discussed in detail in Chapter 4. Iodine is essential for the health of the thyroid gland, which controls the health of the whole body. In cold weather, more iodine is needed as the body's rate of excretion is higher. Therefore, if an animal is iodine deficient, no matter what minerals or vitamins it is given, they will not be assimilated properly until the iodine requirements are met. Any time the cow's system appears not to function properly, an iodine deficiency should be considered as the basis of any problem.

Fortunately, the requirement for iodine is not very high and the addition of seaweed products in the lick is usually enough to meet it. This is preferable to using inorganic iodine in the form of potassium iodide or Lugol's solution. Both are very toxic in excess as well as being expensive. A vet will advise on alternate forms of iodine supplementation if necessary.

Another cause of iodine deficiency is overfeeding of legumes such as alfalfa, clovers, beans, peas, soybeans, lupins, etc. These are termed goitrogenic feeds because, in extreme cases, they cause the goitre to swell. Feeding too many legumes will cause a high ratio of bull calves. The heifer calves have the greatest need for iodine, if there is a deficiency, they may not develop *in utero* and are probably reabsorbed. It is also possible, in some cases, that heifers are not even conceived. Occasionally in iodine-deficient animals strong males are born, but the females are born weak, hairless and dying.

A dairy farmer who milks and breeds stud Jerseys was complaining to me that the cows had produced nothing except bull calves for the last three years. Someone had unfortunately advised him to have clover-dominant pastures and to feed alfalfa without telling him of the pitfalls. Bull calves and ill thrift were the reward for his efforts. The farm was turned around within a year of balancing the soil.

Feeds high in nitrates, like capeweed, when present in large amounts (such as after a drought), and also those high in oxalates, can also inhibit iodine and cause thyroid dysfunction. If animals are on paddocks which are largely growing capeweed or occasionally some varieties of African grasses, iodine supplementation is necessary as well as dolomite. Seaweed meal would usually be enough.

Remember that an iodine deficiency should always be considered as the base cause of practically every problem. A blood test will show a large iodine deficiency, but might not highlight a chronic mild deficiency. It will be found that all stock will help themselves to seaweed meal as they require it when they are in paddock situations, and this as well as the stock lick would be the safest way to provide it.

Excluding the actual sign of swollen goitre (in the middle of the underpart of the neck), one of the simplest warning signs of an iodine deficiency is the presence of dandruff (scurf). Woody tongue is also a condition that can arise when iodine is too low.

The need for iodine is increased by cold weather and, as mentioned, by feeds that are goitrogenic and/or high in oxalates or nitrates. A selenium deficiency can also raise the need for iodine. All this points to the many advantages to be gained from regular supplementation with seaweed products which contain the latter mineral naturally.

Iron

This mineral is necessary for the health of the red blood cells and consequently for the general well being of all animals. Fortunately, as Australia is largely volcanic in origin, iron deficiencies in the soil are very rare. In fact, in this country, the iron content is often so high that it suppresses other minerals and vitamins — potassium and vitamin E in particular. Iron is often the only mineral remaining after years of bad farming. This situation can be easily amended by top-dressing the area with the necessary lime minerals after a soil analysis. Bringing the calcium/magnesium up to the correct level is usually all that is needed.

As mentioned in the section on copper, in spite of the prevalence of iron, anemia would be one of the biggest causes of sickness in stock. This is because iron cannot be assimilated unless enough copper is present in the diet — thus the anemia is not caused by a lack of iron, but generally by a lack of copper.

Iron tonics may be used as a short-term remedy for no more than a week. Copper supplementation will usually raise iron levels very quickly. The disadvantage of iron tonics is that they totally suppress vitamin E. Fish products can be used safely to boost iron levels in an animal if there is a lack of iron in the paddocks. Top-dressing with basic slag, a by-product of the iron smelting industry, used to be done in the past and would possibly be the best way to amend iron shortfalls. However, in Australia, it is not considered to be an allowable input in an organic certificated farm.

Manganese

Deficiencies of this mineral do not appear to affect animals a great deal. However, the research from the United States, as reported in the December 1992 issue of *Acres U.S.A.*, is that a lack of manganese, allied to shortfall of copper, cobalt and iodine, predisposes animals (and humans) to *Brucella abortus* infections (mentioned above). The majority of soils show adequate to good manganese levels. As the soil health is improved, manganese becomes obtainable.

Molybdenum

This element is needed for maximum fertility. A deficiency can occur in land that is exhausted and low in organic matter. This is more likely if the pH is below 5.0.

Far more serious is excess molybdenum which suppresses the uptake of copper. This is a far more frequent occurrence than a deficiency and is probably the reason why, in the United Kingdom, we were taught to adjust the levels of the other elements through top-dressing rather than altering the molybdenum. Increasing the organic matter in the soil is the best remedy. This can be done by aeration, spreading animal manures, composting, slashing the pastures and generally attending to the good health of the paddocks.

Phosphorus

Phosphorus is absolutely essential for healthy growth and life. It should be kept in balance with calcium and magnesium, otherwise an *excess* of phosphorus will lead to bone fragility and many other problems of the kind associated with calcium and magnesium deficiencies. Phosphorus should not be lacking in healthy, well-farmed soils where organic matter has been allowed to accumulate. Soils that have been heavily cropped, and therefore subjected to leaching, will be low in available phosphorus (and everything else).

Unfortunately, excessive use of superphosphate with no reference to the calcium and magnesium balance of the soils has led to too much phosphorus in many cases, causing disease conditions due to a lack of other minerals. Phosphorus deficiencies are not as common as we are led to believe and, if paddocks were properly farmed, would become nonexistent.

In cases where an apparently genuine deficiency is shown in an analysis, have the analyzer give a readout for the unavailable phosphorus as mentioned. On one low-pH farm where the phosphorus levels showed six ppm, the unavailable phosphorus was 600 ppm. This "bank" of unavailable phosphorus is drawn on by adjusting the lime minerals, improving the soil pH and general health.

I am loath to suggest ways of top-dressing with phosphorous, given that it locks up copper. But there are various forms of rock phosphate should it be needed and much of it has the great advantage of being a slow-release mineral. One has to remember that superphosphate really came into being because people no longer wanted to go to the trouble of spreading animal manures on the fields.

Potassium

This mineral is absolutely essential for all life, both plant and animal. It is rapidly being lost worldwide due to chemical fertilizers which raise the sodium level in the soil. Potassium is also inhibited by a soil too high in iron or one that is too acid. On conventional chemical farms potassium is always being lost, whereas it is more than adequately replaced annually on organic farms.

According to Charlotte Auerbach's book, *The Science of Genetics*, a lack of potassium and vitamin C at conception can interfere with the true pattern of inheritance. So when breeding valuable stock, it is essential that the soils be as healthy as possible.

The most serious result of a potassium deficiency is difficult births (dystokia) and having to pull every calf is *not* desirable. This is caused by constriction of the blood vessels to the uterus and cervix in the final stages of pregnancy. When this happens, the correct presentation and muscular contractions do not occur as they should. Raising the health of the paddocks is the best long-term solution. I remember a vet from the Western District of Victoria telling me that in the 1965 drought he had to pull 95 percent of all the calves in his area. There was no green feed and he rightly considered that a potassium deficiency was the cause.

Potassium deficiency also causes an inability to utilize fodder correctly. Unfortunately both muriate and sulfate of potash raise the salt table to undesirable levels. As always, getting the land back to a balanced system is best. Foliar feeding of potash could be used in commercial growing setups and would not be so damaging to the soil.

Selenium

The main reason for increasing selenium shortfalls can probably be traced back to the fact that sulfur deficiencies are the fastest growing deficiencies in the world today. This is according to Neal Kinsey, who told me this late in 1998 and, as his consultancy is global, he would have the figures right. His observations also line up with what I have observed in Australia and England. The bottom line is that without the amino acids of sulfur in the gut, no cow or other species can absorb selenium which is not as scarce as many people think. In parts of the United States it is too high.

A small amount of selenium is needed for fertility, particularly of bulls, and for healthy muscles and growth. Complete and sudden cessation of growth, with or without muscle wasting signs, could be a sign of selenium deficiency in all young animals (also called white muscle disease).

Another sign of a deficiency can be retained afterbirths. Studies in the United States have shown that there is a need for selenium by the fetus in the last stages of pregnancy. Selenium is linked with vitamin E and often giving that vitamin alone will effect a cure in mild deficiency cases. In fact this can be used as a test; if a sick calf responds to 2,000 units of vitamin E, you can be sure that it is suffering from a selenium deficiency. The farmer can then give a seaweed drench for example, or see that the calves have seaweed meal *ad lib*.

All seaweed products contain natural selenium; in this source it is quite the most desirable way of administering it. Richard Passwaters, M.D., in his book *Selenium as Food and Medicine*, states that four grams of sodium selenite are needed to equal one gram of natural selenium as found in seaweed products. Given that sodium selenite is rather toxic, the natural form would be preferable. Excess selenium can cause a malformed fetus and/or poisoning. Sodium selenite (inorganic selenium) is only obtainable through a veterinary surgeon so accidental poisonings should not occur.

Selenium deficiencies are crucial to the stock breeder who runs his own bulls. Lack of selenium causes scarcity of sperm and those that are there tend to drop their tails and be very weak. This is one form of infertility that is, luckily, reversible. Giving bulls about 40 grams of the stock lick at the head of this section, along with a supply of vitamin E, should ensure that this does not happen.

As stated, selenium cannot be assimilated without the amino acids of sulfur, so it is most important that the sulfur in the paddock (raised by using gypsum, calcium sulphate) and in the diet (raised by feeding yellow sulfur) be maintained at the right levels.

A friend who runs about 80 cattle, in this case pedigree Herefords, lost 25 calves one year from selenium deficiency. Fortunately the cattle were watered in troughs, so he added two liters of liquid seaweed per week to each trough — they were the long kind — and the trouble stopped. The need for selenium (like iodine) is increased by cold weather and goitrogenic feeds.

The yucca plant, which grows wild in some places is very high in selenium. I have seen cattle eat it avidly in spite of its extremely hard leaves in order to get what they needed. I have an idea that it is native to the Americas.

Sodium

Salt is essential for life, but in our civilization, there is generally far too much already in the food chain. Too much salt upsets potassium levels, causes edema (fluid retention), cancer and other degenerative conditions. Excess sodium also prevents cattle from using their fodder correctly. Chemical fertilizers and poor irrigation practices have led to a dramatic rise in salt levels.

There is generally enough salt for normal requirements in feeds. Animals needing extra are probably short of other minerals, and adding seaweed in some form or other to the ration would be the best course. Seaweed contains a small amount of natural sodium. In cases where salt is really needed and seaweed products do not supply the requirement, natural rock salt should be offered. I found that properly supplemented stock hardly ever touched salt; new arrivals would take large amounts until their mineral level returned to normal.

Sulfur

In the past, farmers recognized the importance of sulfur and it is often mentioned in old farming manuals. Sulfur is just as important now as it was then, more so in fact because the use of chemical fertilizers means that it is not coming through in the food chain.

Cattle that are sulfur deficient may have lice, ticks or other exterior parasites. They will not digest their feed properly as lack of sulfur interferes with the action of the amino acids, especially cysteine and methionine. Growing animals will not progress as well as they should if sulfur is missing. The mineral is also needed by the calf *in utero* in the last two months of pregnancy.

Sulfur is often beneficial for skin ailments, both topically (applied to the skin) as well as added to the feed; therefore, skin troubles could be another sign of a deficiency. If applied along the back line of cattle with bad infestations of lice, it will give immediate relief and its addition to the diet will stop reinfestation.

Sulfur can be added to fodder and fed to animals at any stage of development. It is completely safe as long as it does not exceed two percent of all feed eaten, which is very unlikely to occur. For lice infestation, a heaped tablespoon every day for a cow until the lice go will be necessary. The stock lick at rates suggested will supply cattle with their requirements and prevent any of these troubles.

Zinc

Zinc is necessary for a healthy reproductive system both in bulls and cows, particularly the former, as the prostate gland has a high zinc requirement. Eczema is caused by a zinc deficiency. Facial eczema in sheep in New Zealand was a considerable problem until it was discovered that it was caused by a lack of zinc. Zinc is also indicated where recovery from sickness is not as fast as it should be, again seaweed products will probably be all that are needed.

Zinc is often missing in the food chain, possibly like other trace minerals because of leaching. Chemical farming was thought for a long time not to affect its availability. However, recent reports now admit that it, like other valuable trace minerals, is also lost to chemical farming.

Zinc and copper were always considered to be antagonistic. It has now been established that zinc does inhibit copper, but the reverse does not happen. Pigs and camelids seem to have an elevated need for zinc, otherwise seaweed products supply enough for other types of stock including cattle. Zinc sulphate can be used for supplementation.

Vitamins and Their Uses

If cattle are receiving the minerals they need, very few vitamins will have to be used. Behind virtually every vitamin deficiency there is a mineral one.

Probably the most important single cause of vitamin deficiencies is a lack of iodine because, if the thyroid gland is not working, the adrenal glands will not work either and consequently vitamins will not be synthesized. If the lick mentioned in the last chapter is available, cattle should be able to take what they need and hopefully will get enough iodine from the seaweed in the lick. Having supplementary licks out for stock is part of the feeding routine worldwide these days and if it contains the necessary minerals, much money could be saved on buying vitamin supplements. Farmers who feed seaweed meal have very few problems with their livestock. In the United States and Europe this is now considered an ongoing and permanent part of the feeding program.

One note before we continue, liquid paraffin should never be used internally on any animal because it destroys vitamins in the gut. If oil is needed for any reason, use a cooking variety that has not been heated. In other words, do not use oil previously used for cooking as it is unhealthy in the extreme.

Vitamins and Their Uses 111

Vitamin A (Retinol)

Vitamin A is a fat-soluble vitamin and is therefore not lost as easily as the water-soluble B vitamins. It is normally stored in the liver in amounts high enough to enable the cattle to cope with a prolonged period on dry feed, approximately three to five months. In a drought which exceeds that time, deficiencies will start to build up.

This vitamin is normally obtained from well grown (minerally balanced and chemical free) green feed. In nature it is frequently bonded with vitamin D, as in fish liver oils, and this is one of the easiest and safest ways of administering it.

Vitamin A is needed for fertility, conception, to prevent reabsorption of the fetus, and genital and urinary tract infections. Cows that cycle normally but return to the bull are usually short of vitamin A. Sometimes a cod liver oil drench before the next estrus will put the situation right. This assumes, of course, that the bull has been receiving his cod liver oil. If he becomes infertile through a lack of vitamin A, it can be irreversible. Bulls must be given vitamins A and D regularly in drought conditions.

Should a female be deficient in vitamin A at calving, her calf will not live beyond nine days as it will not receive enough vitamin A from its mother to sustain life. It is very important therefore to see that female stock has feed rich in vitamin A prior to calving. There is also evidence that calves may be born with contracted tendons if vitamins A and D are short.

A healthy paddock of green feed that has not had artificial fertilizers used on it and has been minerally balanced would be all that is needed. If there has been drought for more than four or five months, it will be necessary to give breeding cows a dose of high potency vitamins A, D and E, either orally or by injection, coming up to calving to ensure that the calves are born healthy and stay alive.

Vitamin A is particularly important for the health of the eyes. An outbreak of pinkeye (conjunctivitis) is a sure sign of a deficiency. The vitamin is needed for the health of the adrenal glands (which in turn depend on the health of the thyroid) and also plays a large part in general resistance to infections. Cattle deficient in vitamin A are at risk from worm infestation. A harsh, dry coat and runny eyes (both signs of worm infestation) can be a sign of vitamin A deficiency (as well as copper deficiency).

Another way in which vitamin A is depleted has recently been reported at length in *Acres U.S.A.*, and that is overhead power lines. This backs up my own finding with goats reared under the main grid. They appeared to have lost the ability to synthesize their own vitamin A. For three years after leaving that locality they needed ongoing vitamin A supplementation in spite of the fact they were living off my organically farmed pastures.

A vitamin A deficiency is also liable to occur in processes where hormones are used such as embryo transplants and similar procedures. Animals who are being used for this purpose should receive vitamin A regularly or they may abort and waste the whole process. Hormones tend to deplete calcium, magnesium and vitamin A.

Vitamin A, like other vitamins and minerals, is deficient (up to 28 percent) in the products of chemical farming. The vitamin is also susceptible to light; if an animal was left in light for 24 hours, its vitamin A reserves would be severely depleted. A period of darkness is needed to preserve the normal functioning of vitamin A. If using a water-soluble vitamin A and D emulsion, do not add it to the water or the vitamin A may be destroyed by the light. Vitamin A is better given by drench as part of the vitamin A and D complex or included in the feed where it will be eaten straight away. (See note at end of section regarding light-proof containers).

If grass hay has no green color at all, one can safely assume it will not have any vitamin A either. Supplementation can be by cod liver oil, vitamin A, D and E powder, emulsion or injections. A cow that has become deficient may require a daily 20 ml dose of vitamins A and D. Too much vitamin A can cause liver damage, but the above dosage can quite safely be given for one or two weeks, as in the case of conjunctivitis, for example (see section on the ailment for specific instructions).

Vitamin A and D injections can occasionally cause trouble at the site of the injection due to the oily nature of the fluid; therefore, the oral route of administration is preferred. The powder (sometimes difficult to obtain) is usually quite palatable and can be mixed with the feed, as can the emulsion or oil.

Of course, the best method of supplementing with vitamin A would be by adding properly grown (on remineralized and chemical-free soils) and harvested alfalfa hay or green feed to the diet if the beast is hand fed. There have been cases where conjunctivitis has been cured by merely moving afflicted stock to a paddock of properly grown green feed.

Vitamin A, and all vitamins, *must* be bought and stored in light-proof containers. As mentioned earlier, light destroys vitamin potency.

Vitamin B complex

This covers a whole range of vitamins: B1, B2, B3, B5, B6, B12, B15 and biotin. Even now, new B vitamins are being added to the list as our knowledge of this field increases.

B vitamins are normally present in all well-grown feeds, especially grains, but they may be missing if the feeds were deficient in magnesium or cobalt as many feeds are. Milling and irradiation destroys B and other vitamins. In theory, all the B vitamins are present in vitamin B complex powders or injections, but in practice it is usually better to give the specific vitamin that is needed for best results.

The product VAM is available as a paste and is an excellent and very effective source of B vitamins. It is a widely available (in Australia) product that contains vitamins, minerals and amino acids in liquid or paste form (see Chapter 11 for dosages).

Vitamin B1 (Thiamine)

Until a few years ago, this vitamin was hardly considered in either human or animal health; if it was, the requirement was thought to be insignificant. We now know that it is absolutely essential and deficiencies are far more common than had previously been supposed. Apparently diets too high in carbohydrates sometimes depress vitamin B1, although I have never seen it.

Deficiency signs can vary from mild distress to staggering, lack of lateral coordination, and blindness followed by death in about 72 hours. An animal deficient in this and other B vitamins often shows scarlet streaking in the membranes of the mouth.

Thiamine is destroyed by thiaminase which is present in some molds, so feed suspected of being moldy should never be fed to stock. The signs mentioned above can be the result. Often this takes the form of photosensitization or staggers. If treatment is not started immediately, death can follow in a day or two. It should be noted that these conditions only appear on pastures that are extremely poor, sour and out of balance. The molds start in the ground and first affect the plants and then the cattle eating the pasture (Dr. William A. Albrecht has explained the process quite graphically in his work). Diseases caused by molds are serious and often cause residual damage.

The dose recommended for this condition is approximately eight mg of injectable vitamin B1 (obtainable from any vet or feed store) per kilogram of body weight intramuscularly every six hours until the signs disappear. One milliliter of injectable vitamin B1 should contain 126 mg — consult the bottle. Vitamin B1, like most B injections, is water soluble. This means that any excess is eliminated by the body and there is not the danger of overdosing as there could be with an oil-soluble injection.

It is important to know that thiaminase poisoning is one of 200-plus diseases caused by molds. Therefore, vitamin B1 may not always work, but it is worth a try, since nothing else seems to work.

Vitamin B2 (Riboflavin)

This vitamin would normally be present in well grown green feed and the need for supplementation would not often arise if animals are well fed. However, it is needed for the health of the mouth and lips and for the eye's ability to withstand bright light. Cataract is sometimes caused by a deficiency. It is also needed in digestion. Deficiency signs could be cracks round the mouth and bright magenta color in the tongue or gums. Difficulty in urinating has also been noted in animals deficient in this vitamin.

Supplementation can be by crushed tablets, but a supply of really well grown green feed should set right a deficiency fairly quickly.

Vitamin B3 (Niacin, Niacinamide, Nicotinamide, Nicotinic Acid)

Any of the above names are used to describe this vitamin, which has only recently been called vitamin B3. Vitamin B3 deficiency is associated with mental trauma like forgetfulness and senility, neither of which is likely to affect cattle. It is, however, an essential part of the B complex.

Vitamin B5 (Pantothenic Acid, Calcium Pantothenate)

Both the above terms are used for this vitamin which plays an important part in all animal's resistance to disease. It is useful in the treatment of any illness caused by an infection. This is because cattle cannot make their own cortisone in the adrenal cortex unless they have enough vitamin B5 as well as vitamin C, which most animals synthesize from their feed. A full-grown bovine makes about 30-plus grams of vitamin C per day in its liver from its food.

The best source of vitamin B5 is barley grown on remineralized fields. Juliette de Bairacli Levy, the great herbalist, emphasizes that barley should always be part of the ration for any animal. In the days when she advocated this, the B5 part of the grain was not known — she merely knew that it kept cows and stock healthy. Pigs in an English commercial setup that were being fed on maize were dying from a condition resembling necrotic enteritis. The deaths also occured when meat meal and alfalfa were added to the ration. Post mortems showed absolutely no B5 in their bodies; only B6, which is contained in maize, was found. The deaths stopped when the maize was replaced with barley.

Ground up tablets of vitamin B5 can be used for treatment of deficiencies in the case of a cow; use about ten times the dose recommended for a human mixed in with the feed. Injections are also sometimes obtainable. Instructions on the bottle should be followed.

Like other B vitamins, vitamin B5 should be found in well grown green feed and grains. Milled grains should be eaten as soon as possible since milling destroys vitamins. Barley can be fed whole, either soaked or dry. All animals soon learn to digest it and it will not be seen in the manure after a few days.

Vitamin B6 (Pyridoxene)

This B vitamin is needed for resistance to infections, particularly herpes-type infections. It is also useful as an adjunct in the treatment of nearly all infectious diseases and should always be considered in the treatment of any illness as it helps other vitamins and minerals work better. It can be obtained in tablet form or occasionally as an injection. As shown above, it is found in maize.

In humans it must always be administered with B complex, but stock should obtain enough of the B complex in their normal feed to make administration of B6 safe on its own. B6 is the antidote for travel sickness in smaller animals.

Vitamin B12 (Cyanocobalamin)

This vitamin is an incredibly important one on any farm (see the section on cobalt in the preceding chapter). It has already been mentioned in the sections on cobalt and iron, both of which are synthesized into vitamin B12 in the gut.

Deficiency signs can range from mild lassitude to serious anemic conditions, nearly always accompanied by a subnormal temperature. Weak calves that have not been able to stand up since birth are often miraculously restored to life by 10 cc of vitamin B12 intramuscularly. Any animal that is off its feed for no apparent reason can respond to the same treatment, 30 cc would be necessary for a cow. Often these symptoms follow the administration of antibiotics and it should be a rule that if antibiotics have to be used, the B12 should be given at the same time to offset the side effects. Loss of appetite is often the main one symptom of antibiotic reaction.

For severe cases of anemia, a course of vitamin B12 would be preferable to the administration of iron tonics with their unfortunate side effects.

Vitamin B15 (Pangamic Acid)

This vitamin is now once again available from veterinarians in Australia. However, veterinarians from the United States that I have worked with had absolutely no doubt about how useful this vitamin is for restoring liver function. It is well worth using in any case of general debility or sickness. Ten to 12 ml for large-bodied cattle or bulls is the usual dosage. It is also a useful tool for copper toxicity should that condition arise, for example, heliotrope poisoning or an overdose of copper sulfate.

I was rung recently about an accident where a large quantity of copper sulfate was spilled into a bucket of molasses. Five racehorses drank the lot and their liver counts went up to 80 (from a norm of around 10). The trainer's vet was from California and she had absolutely no problem with using B15 with the dolomite and sodium ascorbate I suggested. She also added activated charcoal to the mix. All the horses were back in training in five days.

Biotin

This is another B vitamin that has suddenly come into prominence. Except for seaweed meal, it would not occur naturally in any quantity in normal concentrates unless they consisted of whole maize or beans, since any form of milling will remove it from grains. However, cattle would probably not have a great need for this vitamin. (In horses it *is* necessary for feet and hair growth). Cattle regularly fed seaweed products should have no deficiency problems.

Vitamin C (Ascorbate, Ascorbic Acid)

Vitamin C is essential for the health of the cells, strength of blood vessels and for collagen — especially in the pads between the vertebrae and joints. Vitamin C has been used most successfully to cure and alleviate bad backs, particularly in horses when caused by the breakdown of the intervertebral pads (disks). All animals (except humans, some monkeys and guinea pigs) manufacture their own vitamin C in the liver from their feed. A large cow would make about 30 grams-plus a day, but under stress of any sort, be it trauma (being chased for example), sickness, travel, poison bites or whatever, the extra demand may easily exceed the supply.

Vitamin C can be used to control cancer, tetanus, gut infections, viral conditions, blackleg, snake and poison bites, and as an antidote for certain poisons (see Chapter 11). It is a tremendous help in the treatment of any sickness as it helps the body, and the liver in particular, to fight back and recover.

Blood vessel fragility in all stock is helped by vitamin C administration. Hesperidin is a part of the C complex and it would appear to be the factor which helps strengthen blood vessels. Blood in the milk can be caused by fragile blood vessels.

Vitamin C can be given orally in feed in the form of sodium ascorbate powder which is tasteless. It will not curdle milk and is therefore very useful for adding to milk for sick calves. Ascorbic acid powder can be used in feed; it has a very sour taste and most stock seem to like eating it neat. Vitamin C can also be given by injection and is cheap and easily obtainable in 50 ml or 100 ml bottles from any good fodder merchant or chemist. In Australia, a bottle of injectable vitamin C retails at $6.80 (about $4.00 US).

Sodium ascorbate powder (not ascorbic acid) can be dissolved in distilled water for injection. A teaspoon of powder makes about five grams, but be warned that sterile procedures *must* be used. The usual strength for injectible (bottled) intramuscular vitamin C is two cc to a gram; check that it is so when buying it. The dosages referred to in this book are at a concentration of two cc to a gram.

For intramuscular injection either ascorbic acid or sodium ascorbate may be given; but for the intravenous route, sodium ascorbate should be used. As an intramuscular injection, vitamin C can sting slightly, but large animals I have treated do not seem to show this reaction and maybe it is only with smaller animals that this occurs. Blackleg is the exception to this (see Chapter 11).

Vitamin C injections can be used in large amounts with complete safety in situations where an antibiotic might be used. Many vets use it at the same time as an antiobiotic with very satisfactory results. The vitamin has the great advantage that it minimizes the side effects often associated with antibiotic use. Many viral infections do not respond to antibiotics, they are only given to control secondary bacterial infections. But vitamin C, if used in sufficiently large amounts, will cure viral conditions that have not responded to normal drugs.

When used for snakebite, vitamin C has the advantage of working slightly faster than antivenom, without the risk of anaphylactic shock. The variety of snake does not matter *and* the vitamin C is on hand — as near as the fridge. Animals always seem to be bitten by snakes or taken ill at weekends or when the vet is unobtainable, so the vitamin is a useful standby.

Vitamin D (Cholecalciferol)

This vitamin is needed for bone growth and the absorption of calcium. If it is missing, disorders akin to rickets will ensue. In Europe this condition was first noted in children from the slums of the industrial revolution. Their bones did not develop properly and the legs started to curve from the weight of the body. It can and does happen just as readily to young stock. No matter how much dolomite, etc., is added to the ration, without vitamin D, the calcium and magnesium is not correctly utilized. In rickets, cod liver oil (the safest form of vitamins A and D) effects a cure very rapidly.

Signs of vitamin D deficiency are unnatural bone formations and, in mild forms, a very harsh coat. Another sign of vitamin A and D deficiency is young stock born with contracted tendons (knuckleover) in the front legs. Ten ml of cod liver oil straight down the throat, or in the milk (if being poddied), will soon straighten up the legs.

This vitamin should be found in all properly grown green feed, but the chief source of supply is sunlight, from which it is synthesized on the skin. In areas where the sun is strongest, people have naturally dark skin which prevents them absorbing too much vitamin D as it is dangerous in excess. For this reason it should be noted that dark-colored animals could need supplementation with cod liver oil, as they need more sunlight to obtain it than light-colored animals, especially if the year has been cloudy. Animals wearing coverall rugs, like those used for Jerseys, could also need extra cod liver oil.

The vitamin should always be used in the A and D complex as in cod liver oil, which is where it is found in nature, apart from healthily grown feed and sunlight. Used on its own, vitamin D can be very dangerous. Cod liver oil can be bought either as an emulsion, oil, or powder (when it is usually bonded with vitamin E as well). In nature, vitamin D as a food is always found with vitamin A and it is safer to stick to this combination.

Vitamin E (Tocopherol)

This is a fat-soluble vitamin, so it is stored in the body in normal circumstances and it acts as an antioxidant. Unlike many other fat-soluble vitamins, it seems to be safe in large amounts; perhaps the requirement for it is fairly high.

Vitamin E is necessary for general good health and, in particular, is useful in healing after an injury or trauma. Vitamin E is also helpful with vitamin A for fertility problems, especially when allied to a selenium deficiency because vitamin E is in some way bonded with selenium. A researcher in New Zealand, a doctor, found that his pigs, which were artificially raised, suffered from a sudden death syndrome. He performed controlled experiments, supplementing one half of the piglets with vitamin E and selenium, and found they lived, but the unsupplemented ones did not.

Vitamin E should be found naturally in organically grown seeds, unmilled grains, wheatgerm and wheat bran. Fresh wheat germ oil, which should be kept refrigerated, is probably the best source for supplementation if small quantities are needed. Vitamin E is often supplied with vitamins A and D, both in powdered and injectable form and is obtainable on its own. It is the fashion nowadays to make it from soy, this source is not so good as the basic wheat germ-derived vitamin E.

Recent research from a Colorado State University meat scientist, Gary Smith, has found that feeding (not injecting) extra vitamin E, about 400 IU a day per head, prolongs the shelf-life of beef to almost double. Apparently Australian beef is known to have higher levels of this vitamin than its American counterpart. Perhaps our feeding methods include more and better grown *whole* grains and they are a natural source of vitamin E.

Vitamin E is destroyed by mineral oils, such as liquid paraffin which should *never* be used for dosing any animal, but vegetable oils are safe. The biggest destroyer of vitamin E is iron, either tonics, supplements or injections. This should be remembered at all times and, given the information in the section on iron and copper, it will be noted that it is not often necessary to supplement with iron at all in Australia, and possibly not elsewhere either.

Vitamin B12 injections should be used as they do not destroy vitamin E; iron tonics should only be used in the *very* short term.

Vitamin E has been very useful in the aftermath of pneumonia to normalize the breathing rate. It helps to heal the lesions in the lungs, 6,000-8,000 units would be needed daily for a cow.

Vitamin H (PABA, Para-aminobenzoic Acid)

Until recently, this vitamin has been classified as belonging to the B group, but is now referred to as vitamin H. It provides a source of folic acid in the gut and is used by humans to offset the effects of sunburn and sunstroke. Neither condition would normally affect stock, except possibly very white-skinned cattle. Herefords with unpigmented skin around the eyes are often affected by the sun.

PABA is very effective for combatting sunburn and, as usual, about ten times the number of tablets prescribed for human use should be ground up and mixed in the feed when required for a cow. PABA is also involved in the synthesis by microorganisms of folic acid in the gut (see Adelle Davis' *Let's Get Well*.)

Vitamin K (Menadione)

This is a fat-soluble vitamin, so it is normally stored in the body. It is needed for normal coagulation of the blood, and is therefore particularly important for all animals.

Vitamin K is found in grains, beans, wheat germ and good green feed, so it would seem that a deficiency should be rare. However, it is very easily destroyed by mineral oil laxatives, oral antibiotics and some drugs. More importantly, it is totally destroyed by irradiation of foodstuffs. This process is becoming more common these days and there are suggestions that it may be used for grain and some packaged hay (as is sometimes used at shows).

It is best to avoid situations that would destroy the vitamin, but if a deficiency is suspected, supplements of vitamin K have become available in recent times. A vet would be able to advise on quantities. Normally alfalfa should contain adequate vitamin K.

Vitamin K could be tried as an antidote for rat and mouse poisons that destroy the clotting ability of the blood. In such cases, vitamin C would also be useful.

Chapter 10

Herbal and Useful Remedies and Remedies to be Avoided

There are a number of excellent herbal books on the market and Juliette de Bairacli Levy's *Herbal Handbook for Farm and Stable* is one of the best. Mrs. Grieve's *A Mondern Herbal* is another book on which I have relied. The disadvantage in Australia is that many of the herbs are unobtainable and others do not grow in dry conditions. This chapter discusses herbal remedies which I have found to be particularly useful and which are also readily available in my country. There are many more and I encourage anyone who is interested to take up the study of medicinal herbs and plants.

In the United States herbs can be grown in the temperate regions and they are largely available. Their use is more and more widely accepted.

Natural Remedies

Aloe Vera

This is a plant of a succulent family that grows naturally in parts of Australia. Those lucky enough to have the plant often use the leaves directly, otherwise it is obtainable in liquid, ointment or gel form. Care should always be taken to buy the genuine product as there have been adulterated supplies on the market. With any herbal product, it is always best to get as pure a form as possible.

Aloe vera has received a great deal of publicity in recent years and there are many testimonials to its usefulness. It can be given orally or used externally and the latter mode of treatment has produced quite spectacular results in the case of badly ulcerated wounds and burns.

I bought a buck goat some years ago that had foot rot which resulted in a long-standing and very obstinate bone ulcer. Amazingly, aloe vera effected healing in three days. This was after a long struggle of trying anything and everything without success. It could be of use in udder lesions, as the gel dries quickly and helps promote healing.

Apple cider vinegar

This simple and easily obtainable substance has many uses. Cider vinegar contains potassium as well as other trace minerals and helps maintain the correct pH in the body. Feeding a lot of apples as a potassium source could lead to problems in any animal, but they will tolerate cider vinegar in large amounts, and in this form it is wholly beneficial. Because of its potassium content, it is invaluable for cattle just before breeding, especially on potassium-deficient pastures (see Chapter 8).

I first used cider vinegar on my milking goat herd after a season of very difficult births. The next year I was amazed at the difference, even the largest kids from maiden does arrived relatively easily and in very good health. Since then I have used it with my animals at all times, but it appears to be quite effective when used for the last two months coming up to calving.

Cider vinegar helps prevent bruising and assists the tissues in recovery from exertion. Given regularly to stud bulls, it will help prevent stones in the kidneys and ureter, especially useful if your bulls are limited to very hard bore water as is the case on some properties. Adding 500 ml of cider vinegar (about a pint) to the feed twice a week would be sufficient. Prevention is certainly better than cure for this distressing condition.

It can also be used as a mild cure for skin conditions like ringworm. One need only rub it in well on the affected area two or three times a day for a couple of days.

Those wishing to learn more about cider vinegar should read Dr. D.C. Jarvis' very interesting books of various titles on cider vinegar, folk medicine and arthritis.

Arnica Montana

This is a perennial herb that grows in the mountains of Europe and is now being cultivated successfully in Australia. If it can be grown here, it should grow anywhere. Arnica grows best in remineralized soil. It is best used in homeopathic tinctures, pillules, and ointments and is available in health shops. Homeopaths can also provide sprays for badly hurt beasts. In homeopathic form it is an excellent painkiller.

I have used it postoperatively with astonishing results by normal rules. I once used it on a dog who had to have surgery to remove a salivary gland. Fourteen stitches and drainage tubes were required. The dog concerned had no idea she'd had an operation and did not try to scratch or lick the site at all. It seems to have a healing effect as well, the dog in question has no scar from the operation.

In common with vitamin C, *Arnica* is good for shock or trauma. An old book on homeopathy emphasizes that *Arnica* should always be used straight away, no matter what the trauma, as it calms the cow (patient) down and this is very important.

Another case I saw involved an unconscious dog whose owner, trying to stop a fight, hit it with a heavy stick. It recovered within three minutes of placing the *Arnica* under its tongue.

Arnica is available from homeopathic doctors and hopefully, vets as well fairly soon. The 200c potency is most often used for animals.

Comfrey

Comfrey is a broadleaf plant that grows quite readily in the damper, cooler areas. It will not thrive without plenty of water. Unfortunately, comfrey dies back in the winter, but can sometimes be kept going in a sheltered frame where it is protected from frosts.

In spite of much publicity to the contrary, it is completely safe, when taken both internally and externally. In many parts of Germany and Japan, comfrey is used exclusively for dairy cattle fodder during the summer months and they milk very well on it.

Comfrey is one of the few plants that contains natural vitamin B12 which may be one of the reasons why it is so good in the case of sickness. Its old folk name was knit-bone because it helped heal broken bones. Comfrey may be used in poultices and will often reduce swellings in a matter of days. It may be made into an ointment or used as a liquid obtained by boiling the leaves; distilled comfrey oil is the best if available. All forms are useful at some time or other. The plant also has the reputation as an inhibitor of cancer.

The best way to feed comfrey is to offer a few leaves once or twice a week with feed. This is normal practice in many parts of Europe.

Garlic

This is an onion-like plant that will grow very prolifically if kept damp and well fed. Either the bulbs or the chopped leaves may be given. It is also available in oil-filled capsules or tablet form. Garlic, like onions, contains natural sulfur. It can act as a natural antibiotic, especially in intestinal disturbances.

Garlic also has the reputation of being a vermifuge. Although it undoubtedly helps, in my experience here it cannot entirely take the place of good management and attending to the copper levels of the cattle.

In cases of sickness in any stock, persuading them to eat garlic in some form can only be beneficial. It can be blended or offered whole, the farmer will have to experiment. For those involved in dairying, garlic should be used with care as it can taint the milk.

Drugs best avoided if possible

Any student of medicine will know that all drugs have their disadvantages; the question is whether the good the drug does outweighs its ill effects. There is also the fact that we are nearing the end of antibiotics. Animal resistance to them has now become a fact of life and in these circumstances we are fortunate to have vitamin C as an alternative.

Antibiotics

Many years ago the vets from the University of Melbourne tended my dairy goat herd. At that time, goats in Victoria were found with a disease similar to pleuropneumonia in cattle, a mycoplasma infection. The vets used the only drug that they thought had a hope of working, an antibiotic. This was given as a preventative to the whole herd when one of my goats was found sniffing and coughing.

The antibiotic was of the tetracycline family, and a dose that would have been safe for a sheep, an animal very similar to a goat in size and weight, was used. No one knew that goats could only tolerate a fraction of that amount. The result was a disaster to the vets' horror as much as mine. It took the pregnant goats three months to die and the milkers five months because they did not have the same strain on their systems. They all died of acute anemia accompanied by bone marrow damage and renal failure. All efforts to save them were in vain.

From that time on, I was constantly hoping that a safe alternative would be found, and eventually learned about vitamin C and its curative powers. Since that time I have not found it necessary to use anything except minerals and vitamins for any condition that was curable. Vitamin C is the only known substance that can kill a virus as well as other pathogens (at the time of this writing). I have used it for everything from pulpy kidney (enterotoxemia) to tetanus, blackleg and mastitis, usually in conjunction with the required minerals. A book by Drs. Archie Kalokerinos, Glenn Dettman and Ian Dettman states that it has been used to cure foot and mouth disease and the plague.

Now we are a great deal further down the track, and veterinary experience with many different kinds of animals has increased. Hopefully the above account of the destruction of a herd is unlikely to occur again.

Always make sure that a vet gives any drug that your stock may need *after* a sensitivity test has been done to see if the drug will work. *Never* borrow a drug from another farmer or friend, it can be out of date, contaminated if opened, and it has not been checked regarding whether or not it will work on your animal.

Vets tell me that when they give vitamin C with drugs, quite often the good effects are enhanced and the bad ones minimized. Bear in mind the information in the section on vitamin B12, and always insist that an injection of that vitamin is given with an antibiotic if one has to be used (10 cc for cows and possibly 15 cc for a bull). A vet demonstrated this to me years ago with a mare that was very ill. The B12 undoubtedly helped her recover, as it kept her interested in food if nothing else. She ate all I could give her.

Butazolidin (BTZ, Bute)

This would not be given to farm stock in normal circumstances. It is a painkiller that is sometimes used in arthritic conditions, injuries that give pain, or sprains, and often for racing animals. It masks the pain of injury so the animal is unaware of it; not much imagination is needed to see what a very short-sighted policy this can be, especially as the side effects can be fatal. I would not use this substance on my animals.

If the use of "Bute" is ever suggested (as in the case of a show animal), remember that it has one extremely dangerous side effect — it weakens blood vessels, often to the point where internal hemorrhaging will cause death. There have been numerous instances of this occuring. The best course of action is, of course, to cure the cause of the pain, not mask it.

Given the information above in the section on *Arnica*, it should not be necessary to use BTZ at all. Many vets use *Arnica* and other remedies these days.

If BTZ has been used, give vitamin C as soon as possible. The usual dose is 20 cc by intramuscular injection for a full-grown beast for a few days or a tablespoon of vitamin C powder once a day. Vitamin C strengthens blood vessels and so may avert a disaster.

BTZ was found in export cattle carcases which were, of course, rejected. No reason was given for its presence, although there may be an idea that it will act as a growth stimulant. It is *not* a desirable drug and should be avoided.

Cortisone

This is another drug that should be avoided. In humans, cortisone given artificially inhibits the output of natural cortisone for up to two years. I do not think anyone has found out what it does to animals, but I imagine it would be similar.

Cortisone is produced naturally in the adrenal glands in all animals provided the patient is not deficient in vitamins B2, B5 and C. In any condition where cortisone would be indicated, give those two extras, the vitamin C on a regular basis and some VAM once every two weeks, until healing is effected. Recommended dosage is 15 cc of injectable vitamin C intramuscularly and 10 ml of VAM orally (if the injections are not obtainable).

Hormones

Any form of steroids is best avoided if possible, as the after effects are often rather traumatic. I know that drugs and methods of administration have improved over the years, but still have not seen or heard anything that makes me think that one does not pay for the use of such substances. Steroids have the effect of stopping the absorption of calcium and magnesium. Both steroids and hormones also appear to stop the synthesis of vitamin A, so the intake of that vitamin would probably have to be increased as well.

I found this out in large animals that had been given hormones for ovum transplant programs. The animals seemed to have permanent vitamin A deficiency troubles for the next two years. Often the deficiency was so great after hormone administration that they would not cycle and breed naturally for a long period without being given considerable amounts of extra vitamin A. Shortage of that vitamin (and low copper levels) is often the biggest cause of poor breeding performance.

In goats that have been dried off with Stilbestrol, the same has been the case. Presumably no one would be unwise enough to do this with cattle. It took a whole lactation before their milk production returned to normal, and six months to cure the damage to the udder. After all, Stilbestrol is a male hormone which would be contraindicated in a milking animal. Drying off animals with Stilbestrol is also counterproductive because it acts too quickly and often causes mastitis.

Ovum and embryo transplants are our most useful tool for gaining genetic diversity and breeding up new lines. As long as the farmer realizes the pitfalls involved, there should be no problems with their use.

BGH (Bovine Growth Hormone), which is being used by some dairies in the United States, does not seem to be very logical either. Reports suggest that cattle on farms that have been remineralized and are in top health organically are producing more and healthier milk than that produced by BGH-dosed cows.

With BGH the cattle do give more milk (quality not specified), but the attendant rise in mastitis and other ailments has been so high that there is a real danger of drug contamination in the milk. Also vet fees are proving prohibitive. Until the use of this hormone was officially approved by Congress, the company making the drug was not bound to produce information on any side effects. When it was approved for use, *Acres U.S.A.* ran a listing of the adverse effects. The trade name is omitted in this copy (but not in the *Acres U.S.A.* article), in the following it is referred to only as "the drug."

Reproduction. Use of the drug may result in reduced pregnancy rates in injected cows and an increase in days open for first calf heifers.

Use of the drug has also been associated with increases in cystic ovaries and disorders of the uterus during the treatment period. Cows injected with the drug may have small decreases in gestation length and birth weight of calves and they may have increased twinning rates.

The incidence of retained placenta may be higher following subsequent calving.

Use of the drug should be preceded by implementation of a comprehensive and ongoing herd reproductive health program.

Mastitis. Cows injected with the drug are at an increased risk for clinical mastitis (visibly abnormal milk). The number of cows affected with clinical mastitis and the number of cases per cow may increase.

In addition, the risks of subclinical mastitis (milk not visibly abnormal) is increased.

In some herds use of the drug has been associated with increases in somatic cell counts.

Mastitis management practices should be thoroughly evaluated prior to initiating the use of the drug.

Cows injected with the drug may experience increased body temperature unrelated to illness. To minimize this effect, take appropriate measures during periods of environmental temperature to reduce heat stress.

Use of the drug may result in an increase in digestive disorders such as indigestion, bloat and diarrhea.

There may be an increase in the number of cows experiencing periods of "off feed" (reduced feed intake) during use of the drug.

Studies indicated that the cows injected with the drug had increased numbers of enlarged hocks and lesions (e.g., lacerations, enlargements, callouses) of the knee (carpal region).

Second lactation or older cows had more disorders of the foot region. However results of these studies did not indicate that the drug increased lameness.

Injection Site Reaction. A mild transient swelling of 3-5 cm (1-2 inches) in diameter may occur at the injection site about three days after injection and may persist up to six weeks following injection. Some cows may experience swellings up to 10cm (4 inches) in diameter that remain permanent but are not associated with animal health problems. However, if permanent blemishes are objectionable to the user, administration to the particular animal should be discontinued.

Use of the drug in cows in which injection sites swellings repeatedly open and drain should be discontinued.

Additional Veterinary Information. Care should be taken to differentiate between increased body temperature due to use of the drug from an increased body temperature due to illness.

Use of the drug has been associated with reductions in hemoglobin and hematocrit values during treatment.

I do not think any comment is really necessary; it sounds as though the vet would virtually take up residence on farms that used this very doubtful technology. It is interesting to note that nothing has been said about the quality of the milk.

Ivermectin group of drugs

This group of drugs, of which there are a number bearing names generally including the letters MEC and in which Ivermectin is the base ingredient, are incredibly powerful. They were planned to be the ultimate anthel - mintic (worm drench) and parasite (external) control. It was also claimed that there could be no drench resistance to the drug. This has already been disproved in a number of countries, including Australia. All chemical drenches end with animal resistance, which is why prevention through good pasture and animal health along with natural remedies is so critical. The drug has the effect of killing all the unwanted parasites, internal or external, but unfortunately it cannot differentiate between good and bad, so it tends to wipe out the beneficial gut flora with the worms.

I have the original printouts of guidelines from the United Kingdom, and they state categorically that this drug cannot be used on lactating animals, otherwise the milk is not to be used again during that lactation. It is even suggested that it should not be given to calves. These guidelines have not apparently been taken very seriously here.

This group of drugs has, in a number of species, caused sudden death or illness, possibly due to the severity of its action. If it *has* to be used, give it with your vet's assistance. If the animal is still lactating (unlikely in a state that requires treatment with this drug), do not use the milk again that lactation. Also be sure to give the patient vitamin B12 by injection daily for two or three days afterwards to help re-establish its gut flora.

Perhaps the most unfortunate effect of this drug is that the manure from Ivermectin-treated stock is never absorbed into the soil or processed by soil fauna. It is, in effect, poisonous. Of course, the drug could not be used in an organic setup.

The information already given on copper and worms should make this drug obsolete. See Chapter 11 for management alternatives that discourage worm burdens.

Vaccinations

These are a grey area. Vaccines were first invented by Louis Pasteur, who was a contemporary (and junior) of Antoine Bechamp. Bechamp's book, *The Blood,* points out that germs, bacteria or organisms change their form as the host's diet and circumstances change. This is putting it very simply. What it means, in effect, is that a vaccine, once suited to a particular kind of germ (rather like pieces of a jigsaw puzzle) will continue to work until the circumstances of the host change. It can be a change in diet or surroundings; in any case, the vaccine (or immunization) loses its efficiency.

Tetanus vaccine particularly is not considered in many circles to be as efficient as it use to be. One important fact *must* be noted and remembered. Any vaccine or immunization has the action of depleting the patient's vitamin C reserves, sometimes to the point of death. This has been recorded in human and animal circles (see *Every Second Child*, by Dr. A. Kalokerinos). Therefore, any animal due to have a vaccination whose vitamin C reserves could be low, should be supplemented with at least 20 cc of the vitamin prior to receiving the vaccine.

Two years ago, a horse was brought up by some neighbors from a very starved, dry area. I noticed it looking very ill and went across the road to ask if they had any snakes around and they had. I went back home, drew up 20 cc of vitamin C in the syringe, returned and injected the horse. I was not sure that it was snakebite, as the animal's eyes were half shut and it was collapsing in the back legs by that time. Recovery was almost instantaneous, and I could see that its eyes were normal and the pupils not enlarged as they would have been if it was a snake.

The owner said it can't be anything else as the person they got it from had vaccinated it for strangles and tetanus in the float as it was leaving. They were lucky; the animal was saved. This is just one example of why vaccines must be used with caution and carefully monitored.

Chapter 11

Common Ailments and Remedies

The purpose of this chapter is to help farmers be responsible for the health of their cattle. With experience and knowledge of cattle ailments and remedies, one should need to seek veterinary treatment only for really serious conditions. In the event of serious illness, it is always good to be doing something to help the beast while the vet is on the way. Additionally, illness always seems to strike at inconvenient times, such as weekends and holidays, when the vet is unavailable. But often, with a bit of knowledge, you can care for your own animals.

Prevention is *always* better than cure and knowing your animals and their behavior will go a long way toward warding off illness. The alert farmer realizes when animals are off color before they show definite signs. The cow that kicks unusually when the cups are being put on should always be regarded with suspicion; her udder is likely causing her trouble. Similarly, the beast that lies away from the others and is slow to go out or come in should be watched. It is no good to wait until they are down with their legs in the air and then expect a hard-worked vet to pick up the pieces. The vets with whom I worked in the early days used to complain that I spotted an incipiently ill animal so early that it was cured before they found out what was the matter — surely a desirable way to go.

When an animal is sick, sensible care and attention — keeping it sheltered, quiet, well fed and watered — is all important. Good nursing has helped many an animal survive that had no apparent hope of living. The more you are able to diagnose and care for your animals early on, the less illness and trauma they will suffer overall.

Bovine Vital Statistics	
Centigrade Temperature:	38.5
Respiration:	10 - 30 per minute
Pulse:	60 - 80 per minute
Gestation:	282 days
Cycle:	21 days
Cycle length:	16 - 18 days
Ovulation:	12 - 42 days
Weaning:	2 plus months

Abscess (Cheesy Gland, Grass Seed Abscess, *Caseous Lymphadenitis*)

Strictly speaking, all these are not caused by the same thing, but where there is an infective agent the causative germ is usually corynebacteria. It is virtually impossible to stop one of these abscesses from coming up. Drugs do not help and, indeed, if the abscess is lanced before it is "ripe", the resulting sore mess will make the operator remember not to do it again. If lanced too early, it will form another abscess and the process goes on for weeks.

The best approach is to catch the boil just as it is ready to burst. Wear rubber gloves and wipe out the pus, burning everything used for this process afterwards. Once the abscess is clear, syringe it out well with a mixture of two tablespoons of copper sulphate, one tablespoon of vinegar and a pint of water. Put a good antiseptic cream well into the hole, healing will be effected in a day or two and the abscess will not get fly blown. Flint's Medicated Oil is also very effective. This is an old country remedy that was known in the United Kingdom as green oils; I do not know if it is available in the United States.

An injection of five to eight grams of vitamin C for a small cow may be given when the abscess is forming as this sometimes will hurry the process along. In cases where the abscess is on the side of the throat and is very big, occasionally it will burst through to the gullet. When this happens the poison goes into the system. Injections of vitamin C as above, *and* an oral dose of a tablespoon a day of vitamin C powder, should be given for several days to stop the risk of further infection internally.

Caseous lymphadenitis (CLA) is the name given to a condition where the boils become endemic. They do not limit themselves to one abscess, which is often caused by a grass seed, but continue down the line of the lymph system. If the animal is really healthy, its immune system can sort it out, otherwise the boils often end up forming inside the animal, generally with fatal results.

CLA has been classified in the United Kingdom as a zoonose (meaning it can be caught by a human). Use hygienic processes when dealing with *any* boils.

Acetonemia (Ketosis)

This is a problem in dairy herds, especially those run intensively for high milk production. Signs can be a preference for hay and coarse feed, a sweet smell on the breath and milk, and excessive licking and chewing. This condition does not occur in cattle who are fed a diet where the carbohydrates and protein are in the right balance. They must, of course, receive their correct amount of minerals (see Chapter 8). Work done in the United States suggests that the copper in the lick is all important in preventing ketosis.

Years ago a dairy farmer asked me if I had any bright ideas on how to deal with acetonemia and I suggested he add a tablespoon of dolomite per head per feed. The next time we met he was delighted to tell me that it had worked and that there had not been any new cases since he started using it. Again, the lick would be a preventative, as would lowering the protein in the diet (see Chapter 6).

Anemia

This is a big problem in Australia. Signs are ill thrift, wasting and pale membranes, particularly on the inside of the bottom eyelid. If the cows have been receiving the lick, the worms mentioned here should not be a problem.

In Australia anemia is not generally caused by a lack of iron, there is usually too much of it. But iron cannot be used without copper and a lack of that mineral is generally the cause of iron anaemia. Indirectly this can be the reason for infestations of blood-sucking worms such as Barber's Pole worm (*Haemonchus contortus*) and Brown Stomach worm (*Ostertagia*)

Lack of cobalt can be another cause of anemia. Get a soil audit done and see if colbalt is the problem. If it is, immediately add some cobalt sulfate, about half a pound, to the standard lick. If the anemia is due to a lack of cobalt, in which case the membranes may be quite a good color, injected vitamin B12 and cobalt supplementation will be needed. Signs of cobalt anemia are sub-normal temperature and cold extremities. Scouring eventually ensues, as will death if something is not done.

The reason for the anemia must be removed; worms must be killed and the red blood count built up again. Raising the amount of the lick in the diet should achieve both objects. Iron tonics must not be used for more than a week because they depress vitamin E. Daily vitamin B12 injections and VAM orally will be a great help. Both of these should be given in a daily 10 cc dose for a period of one week for a cow.

Anesthetics

Cattle, and most animals, have a higher pain threshold than humans and can stand more than we can, so do not equate the pain of marking a calf with yourself. They recover remarkably quickly and do not do any better if anesthetized — in fact they do worse. Anesthetics are very hard on animals (and people). Nonetheless, people can bear pain better than animals because they know what it is; animals in pain go down very quickly if appropriate measures are not taken to relieve them.

As mentioned above, anesthetics are very hard on an animal. If at all possible a local anesthetic is always preferable to general anesthetics. However, if a general anesthesia has to performed on a valuable animal, see that it is given an intravenous shot of at least 30 grams of vitamin C before the anesthetic. This ensures that the beast does not struggle when it wakes up. It merely comes around as though it had been asleep. In large animals, the struggles that almost invariably follow general anesthesia usually cause more damage than the reason for the anesthetic in the first place. I helped a vet with an operation when he first used the method of giving a large dose of vitamin C prior to anesthesia. He was most impressed by the post-operative difference and how quickly the animal recovered.

Arthritis

Systemic arthritis is caused by malabsorption or lack of calcium, magnesium, copper and boron. Consult the sections on calcium and magnesium and the requirements for absorption. An animal receiving the lick and *ad lib* seaweed meal would be at little risk as long as it was not being fed a diet too high in phosphates or protein. Read Chapter 6 to learn more about getting and keeping the diet in balance.

In some districts where boron is totally missing from the soil, a little extra should be given to an arthritis sufferer. A small teaspoon of Borax per day for a cow for the first two weeks of treatment, after that twice a week in its feed, is usually sufficient. There is natural boron in seaweed, but that may not be enough for stock if the soil is lacking. The soil analysis would have showed this; hence the importance of having it done. Knowing your soils is one of the best things you can do to prevent shortfalls of nutritional elements.

Cider vinegar added to the diet also helps with arthritis. The feed should be confined to good grass hay, chaff and bran with a minimum of grain. As above, adjust the high protein foods in the diet until the cow is recovered.

Arthritis, Infectious (Navel Ill)

This very often starts from an organism contracted via the navel cord. It can also be caused venereally if the male animal has served a female with an infection in the uterus. The organism can then be passed to the next female he serves, or the male may contract arthritis himself. There is also the unlikely event of it being contracted from a wound.

The only way to prevent navel arthritis is to disinfect the navel cord with alcohol when the calf is born, especially if birthing occurs in an old yard or shed. Methylated spirits or iodine will also do as disinfectants. Making sure that the cows drop their young in clean, uncontaminated paddocks will go a long way toward preventing navel ill. Old sheep yards, etc., are *not* good since many pathogens live happily there waiting to infect the next beast that comes along.

If the male is a stud animal, refuse services to females with doubtful breeding history or who are not completely healthy, unless a vet's certificate showing a clean vaginal swab is produced. Females can only be swabbed when in season.

Infective arthritis (which is generally corynebacteria) is a very difficult organism to treat since its presence often is not discovered until the animal is very ill. This is quite a while after the actual infection takes place, which by then will have gained a good hold. The signs are similar to ordinary arthritis. The joints are hot and swollen, but this kind is usually accompanied by a high temperature and misery.

Large doses of vitamin C, preferably intravenous, could be tried along with supportive measures such as vitamin B12 injections and VAM A calf can be given 15 grams of vitamin C intravenously for the first dose, 10 grams thereafter daily until improvement takes place. When the animal is improved, the vitamin C injections can be changed to oral administration of about 10 grams daily. I have used this method successfully with horses; a vet and I did the work. However, it did not repair the damage to the joints. As this organism causes irreversible damage to the inside of joints, treatment with conventional drugs is rarely successful. In fact, if the infection is severe, nothing seems to work very well, so if the illness has been of any duration, it is kinder to put the animal down.

Artificial Colostrum

This can be made with cod liver oil and straight liquid seaweed (containing no extra chelated minerals or urea) added to milk. A tablespoon of cod liver oil and two tablespoons of the seaweed should be given in a liter of milk. While this, of course, does not contain the antibodies which only the actual mother of the calf can give, it does provide a laxative effect so the meconium (beastings) is voided. It also gives the calf some extra vitamins A and D in a natural form. The seaweed provides a broad spectrum of needed trace minerals. An injection of vitamin B12 (10 cc) should be given intramuscularly if the calf shows signs of weakness or shock; it is an excellent booster, as is oral VAM. It is always worth giving 10 cc of vitamin C by intramuscular injection (in the same syringe as the B12) as this helps in shock and should forestall any infections until the calf is on its feet.

Avitaminosis

This condition literally means that the cow/calf has run out of essential minerals rather suddenly. Unusual lethargy, unwillingness to move, eat or drink are the first signs of this ailment. Examine the membranes of the mouth, they will either be streaked with scarlet lines or be a bright pillar box red all over according to the severity of the condition. Give the affected beast *ad lib* stock lick, extra vitamin C, 100 ml of liquid seaweed and/or give it unlimited access to seaweed meal as well. Usually this is enough, but in a severe case, the treatment may be repeated eight hours later. I have seen this condition in goats and horses. One goat I had ate nearly four pounds of a mineral supplement without stopping and recovered almost instantaneously.

Note: Seaweed liquid should not contain added molasses as this encourages biting insects.

Blackleg

This a Clostridial disease and is reported in *Hungerford's Diseases of Livestock* to be incurable. It is not incurable, but large amounts of injectable vitamin C are needed. The disease is caused by *Clostridium Feseri*, *Bacillus Chauvoei* and *B. anthracis*.

It has two stages. First a leg, usually a back one, swells to such large proportions that it sticks into the air when the cow is lying down. Then, if no action is taken, the swelling will spread and the animal will die very soon from the enormous pressure of the swollen parts which rupture and turn black, giving the disease its name. The disease is caused by the organism gaining entry from a wound. The administration of five-in-one vaccine is supposed to control the disease, but several cases have been known in vaccinated stock following an untreated wound.

This is one illness where the whole body is super sensitive and an injection of 50 grams (100 cc) of vitamin C should be given immediately directly into the affected leg, where it does not seem to hurt. Repeat it in 20 minutes. By this time the swelling should be starting to resolve. Once the leg can be rested on the ground, continue the rapid injections for another hour and then, hopefully, vitamin C in food can be offered.

Be prepared for the treatment to take ten days, but it could take less. Vitamin C seems to be particularly effective in clostridial diseases and there would be no harm in giving extra. Good nursing should bring the animal through; if the site of the wound can be seen, disinfect with the copper wash previously mentioned (two tablespoons copper sulfate, one tablespoon vinegar and one pint of water).

Bloat

Bloat is sign of a sick farm, the cause being an imbalance of potassium, magnesium and sulfur or it can follow a top-dressing with artificial fertilizers. If the land has been farmed organically and remineralized it will not occur. The stands of solid clover that so often cause bloat only grow on unbalanced over-fertilized (using artificials) and under-mineralized soil.

If an animal is only mildly bloated, a drench of about a liter of cooking oil (not liquid paraffin) will help lubricate the insides so some of the wind can be dissipated from one end or the other. The oil drench should be followed by enforced exercise. Then another drench should be given consisting of a tablespoon of dolomite and the same of seaweed in about half a liter of cider vinegar (do not try to put that mixture through a drenching gun, shake it up in a bottle and pour it in).

If the bloat is acute and the animal is down, it will be necessary to release the gas. If this is not done, the pressure will build up to the point where the beast suffocates and/or the organs cannot function. The gas is released with a pointed knife or a trocar; the latter is a sharp, hollow instrument that allows the gas to escape. The gut should be pierced on the left side about a hands-width behind the last rib, halfway down the side. If using a sharp knife, insert it and twist slightly, the gas will come out very fast. Only put the knife or instrument in just as far as is needed to release the pressure. Be sure to disinfect the opening before and afterwards. Another drench of seaweed meal and cider vinegar will help the animal recover. An injection of vitamin C would also be a good safeguard (about 20 cc).

Blood in the Milk

This is usually caused by a blood vessel breaking in the udder, often at the start of a lactation due to the cow's milk coming down very fast. It can also be caused by a blow from a horn, in which case there may be a mark, or it may be the precursor to mastitis. Ease off on the grain and give the cow a tablespoon of vitamin C in the feed and the same of dolomite as well as its normal amount of the lick for a couple of days. It usually clears up very quickly. Vitamin C strengthens blood vessels and helps recovery from injuries.

Bovine Ephemeral Fever, *see Three Day Sickness*

Bovine Spongiform Encephalopathy (BSE, Mad Cow Disease)

This is the bovine version of scrapie, the sheep disease. It is caused by a prion organism, which contains protein and can reproduce, but has no DNA or RNA. There is known cure at the time of writing. Scrapie has been known in sheep and goats since it was first documented 250 years ago and, up to recent times it has, as far as is known, stayed within those species.

It is believed that the disease is passed to cows through the feeding of meat meal, thereby forcing the herbivore to become a cannibal. In the early 1970s, the low heat (and cheaper) method of making meat meal was adopted, apparently worldwide. It certainly was in Australia which managed somehow to escape the contamination. A few years later, the first cases of BSE and its human counterpart became evident and the rest is history. Now there is even more concern because it has crossed into other species in various forms. There are also two similar illnesses that have affected humans (which do not concern this book). In animals it causes, besides mad cow disease, a chronic wasting disease in deer, transmissable mink encephalopathy in mink and PSE, mad cat disease. All forms are fatal.

In the United Kingdom some people were affected after eating meat from diseased cows before the presence of BSE was realized. This fact has nearly wrecked the British beef industry. According to press reports in the *Australian* of March 21, 1994, when it was first discovered it was dubbed "Bovine AIDS" because it often takes years to develop the "mad cow" symptoms. It was assumed it would go away. It did not. It was identified on 20,000 British farms and 130,000-160,000 affected cattle have been slaughtered. Brought up as I was in the middle of the last century, the idea of turning herbivores into cannibals was unthinkable. The old people would have said: "You'll pay." We did.

All cases of the disease were caused by feeding the remains of diseased animals in the form of meat meal. Scrapie is probably endemic in species in Europe, which is why sheep and goats are not permitted to be imported to any other countries without *very* stringent quarantine requirements. The meat meal that caused the transmission to the animals concerned contained the affected brains. Prions are very difficult to denature, their biological activity can only be killed by prolonged boiling in very powerful chemical detergents and it is thought that extremely high doses of radiation may do the same.

There is really no excuse for taking chances on this one. Meat meal and animal waste, whatever the source, must *not* be fed to stock. In 1990, 5,000 feedlot cattle died of botulism on two separate Queensland feedlots as a result of being fed processed chicken manure. The only reason for feeding meat meal and similar materials in the feed is to raise the protein levels. Due to the degradation of much of our cropping land, proteins in grain are now extremely low. They used to be from 15 to 20 percent thirty or more years ago; I heard of only five and six percent in one crop in the late 1990s. However, properly fed lupins as an additive are a reasonable form of extra protein if required. Of course, barley and other grains grown on remineralized and organically farmed soils contain good sources of protein.

(Prusiner's original thesis on the scourge appeared in the October 1984 issue of *Scientific American*.)

Brucellosis (Bang's Disease, Undulant Fever)

The latest research from the United States links this disease with diets deficient in iodine, cobalt, copper and manganese. All four of these together have been helpful in reversing the condition. Again prevention is better than cure. There is not as much brucellosis about as there used to be, but it is by no means eradicated.

Buffel Head (Nutritional Secondary Hyperthyroidism)

This condition, which is caused by high oxalate content in African-type grasses, is thought to only affect horses. This it does by depleting calcium, magnesium and iodine to the point where they die. Their heads and other joints swell up and they go down fairly quickly.

Recently, however, I have been rung about a stud buck goat and a stud bull who were affected. In both cases, the animals had been grazing heavily on African grasses (the bull on Guinea grass). Supplementary dolomite and extra seaweed meal completely averted the condition in horses thus fed, and reversed the condition in the animals mentioned above. This has already been discussed in Chapter 6.

Bulldog Syndrome

This condition usually occurs in young cattle. The animal has no strength in its front leg and shoulder joints. I have only had this condition described to me, but I feel certain that it must be due to a diet lacking in the essential bone minerals. Treatment that includes a dessertspoon of dolomite twice daily, the same of vitamin C, seaweed products (either six ml of concentrate or a dessertspoon of seaweed meal twice daily), and a half teaspoon of boron (this for three days) could be beneficial. It also would be worth giving the animal the basic lick. Prevention, if it can be found, is always better than cure.

Cancer

In stock this is nearly always due to a diet too high in phosphate-rich food from paddocks that have been over fertilized with superphosphate and therefore lack copper, magnesium, calcium and potassium. Lack of iodine and vitamin A can also be causative factors.

If the animal is very valuable, megadoses of vitamin C, accompanied by extra vitamin A, will help. Use 100,000 units of vitamin A per day per cow for two or three weeks and 50 grams of vitamin C by injection daily for a week, with an additional two to four tablespoons by mouth. Make sure the animal has organically grown food and no grain. Avoiding all legumes as much as possible will also help.

André Voisin claimed that cancer is mainly due to a lack of copper in the food chain. Remineralized pastures and the basic lick should prevent it. Care also must be taken that the cows are not in contact with poison sprays of *any* kind.

Coast Disease

This is the name that was given to a wasting condition that usually affected calves in coastal areas of Australia and it could occur in cobalt-deficient areas of the United States. The wasting is caused by a lack of cobalt in the soil that is possibly endemic or perhaps caused by leaching and/or chemical fertilizers. The remedy in the short-term is to give 10 cc of vitamin B12 in the highest potency obtainable by intramuscular injection. The long-term and the most satisfactory remedy is to incorporate cobalt sulfate into the licks; one pound would be a starting point. Cobalt, like many trace minerals, is highly toxic in excess and fatal in deficiency. Remineralizing and restoring the health of the paddocks to a pH between 6.5 and 6.8 usually means that the cobalt becomes available again. If not, at that pH it can be put on with top-dressing.

Coccidiosis, *see Worms*

Corkscrew Penis (PSDP, Premature Spiral Deviation of the Penis)

This is as the name implies and researchers at Murdoch University in Western Australia are considering that it is possibly a hereditary condition. In large herds, the loss of fertility from this cause is very expensive because it may not be discovered in time. It is caused by weak dorsal ligament in the penis.

There is, however, a possibility that it is another condition which has arisen due to a calcium/magnesium deficiency. Certainly weak ligaments in horses can be due to this cause. Putting out the mineral licks for the bulls would possibly be an inexpensive way of solving the problem. Many conditions are labeled hereditary, but a great many often turn out to be environmental.

Cowpox

This generally affects animals in their first lactation and is very contagious among animals at risk. It is caused by a herpes-linked organism and, if nothing is done, is difficult to stop before it runs its cycle (three weeks). This condition only strikes when the cow is deficient in copper and an exterior copper/cider vinegar wash, as used for scabby mouth or ringworm, helps the scabs to dry up and drop off.

In very deficient animals cowpox can spread over the whole body and care must be taken to see the sores do not become infected. Vitamin C injections should be given before that occurs and the stock lick should be freely available or mixed in the feed at a rate of 30-plus grams per day. See that the cows are receiving their vitamins A and D, as in cod liver oil, regularly as well.

Depraved Appetite, *see Hardware Disease*

Dermatitis, Pustular

This is similar in appearance to cowpox and the same wash will give relief, but extra vitamin A and the lick should be given internally as well. Seaweed meal might be all that is needed. Some of the very high powered cures available these days are so severe they can make the condition worse.

Diarrhea

This is caused by an imbalance in the gut due to poor feed, lack of minerals, or interior parasites, all of which can place the cow at risk. However, Tom Hungerford, the basic veterinary authority in Australia, suggests that diarrhea is nearly always due to a shortfall in copper. Give the lick by mouth — just put the powder straight in. Care must be taken that the patient does not dehydrate. Drench in liquids if necessary.

Sometimes a tablespoon of vitamin C and the same of dolomite works well; half this amount is very good in calf scours which is usually caused by a lack of magnesium in the diet. However, as Hungerford states in his *Diseases of Livestock* that a lack of copper is often the cause in weaners and adults; it is also often the cause of worm infestations as well.

A beef cattle farmer I knew had 80 head that were in a bad way. He rang me because one was down and he had tried every drench in the book without success. He brought the ill one into the cattle yards with the tractor. I suggested that he give it two tablespoons of the lick and the same of vitamin C morning and night for two days. I said that by then it should be well on the way to recovery. He told me that it jumped out of pen the day after that. He then ran the remaining cattle through the race and gave each their two tablespoons of the lick. I asked him if it was difficult. He said he opened their mouths with his left hand and with a scoop that held the exact amount of the dose, he threw it into each beast's mouth. The job took him just over half an hour and the herd recovered completely. Obviously the soil health had to be attended to and the lick made available *ad lib* at all times. The lick must be kept dry or the copper is lost by chemical action in half an hour.

Dystocia

This is the clinical name for difficult births. They are due to a potassium deficiency which causes constriction of the blood vessels to the cervix and uterus. Organically grown pastures are a safeguard; potassium should be readily obtainable in them. As a short-term preventative, add cider vinegar to the cows' rations for a month or two coming up to calving. It can be added to feed at the rate of about a quarter of a pint three times a week or watered onto hay. This is far less trouble than pulling calves. The difference to the calving process when it is used is very marked.

Eczema

This condition is usually shown by scabby and suppurating areas. It is not infectious, like mange, which it resembles slightly. The eczema may occur on several animals at once, but that is because they are all similarly affected by the deficiency of zinc.

Eczema is not so common in farm animals as it appears to be in pets, which are often fed grossly inappropriate diets. A teaspoon of zinc sulfate a day should be given to a cow that has eczema (not dermatitis, that is caused by a copper deficiency). If this becomes a frequent occurrence, check on the soil analysis and add one pound of zinc sulfate to the lick mixture.

Enterotoxemia (Pulpy Kidney)

Enterotoxemia is caused by *Clostridium welchii*, or in rare cases, *Clostridium perfringens*. Both organisms normally can be found in the gut of any ruminant. If the diet becomes unbalanced due to worms, inadequate minerals, lack of iodine, and/or minerally unbalanced paddocks, the bacteria starts to proliferate and as they proliferate they produce a deadly toxin. Death is usually rapid.

Unfortunately, the vaccinations which are often considered to be infallible do not work if conditions and husbandry are bad enough. The vaccinations for this ailment are usually included with those for tetanus or in other combinations (two-in-one, five-in-one, etc.). Normally animals have boosters each year although it is the young who are most at risk.

Many vets have told me that they are beginning to feel that the dangers of enterotoxemia are often overrated, and that if husbandry and land management particularly are good, it is not the scourge we have been led to believe. Personally, I have never vaccinated my herd of milking goats (though I've endured some scorn because of it) and found it was not necessary. Enterotoxemia is an infection which strikes animals when they are below optimum health from some other cause.

Massive doses of vitamin C by injection every few hours, at least 30 cc for a cow, and drenches of dolomite and vitamin C powder could possibly work. The antitoxin is fairly effective, but I have seen a spate of birth defects after using it, which never occurred before or since. Vaccinations are not the answer in many illnesses — good management is.

Flag

Flag is a condition which occurs when a cow first comes into milk. The udder is hard and the milk does not let down. Flag is usually worse around the sides of the quarters. Removing all legumes from the feed will go a long way to helping it clear up and even clover should be avoided. Feed the lick and yellow feed, i.e., chaff, bran, grass hay, etc. The condition improves very quickly if this done.

Flag definitely seems to run in families. I have noticed that it is usually the highest milkers that are affected. Once the condition clears up after a few days the udder is quite unaffected.

Foot and Mouth Disease

This is a notifiable disease. Symptoms are the appearance of small runny vesicles around the tongue, lips and between the toes of ruminants. It is spread by birds, infected feedstuffs and direct contact. Cattle do not eat and become very lame.

An authenticated story from Holland told how two out of three farms on the same neck of land had foot and mouth. The cows on the middle farm, which had contact with the other two, were organically farmed and fed seaweed meal *ad lib*. They never did contract it.

Holland, unlike England and Australia, did not have the total eradication policy, so the above situation could not arise elsewhere. Under a total eradication policy, all cows on the three farms would have been destroyed regardless of their genetic or commercial potential.

There are other conditions that look like foot and mouth disease so always check with a vet if there is any doubt. It has been recorded in Australia several times since the first outbreak in 1870, though it has never spread as it does in cold, wet countries. All recent scares in this country have been unfounded.

Foot Rot

This is another highly contagious disease in animals at risk through copper deficiency. The organism lives in most pastures and copper deficient animals will very soon pick it up. The winter and spring of 1992, which was incredibly wet, produced an amazing number of calls from people with foot rot afflicted stock, and an equal number of thankful ones who, when supplementing with the stock licks, had cleared it up very quickly. One lady who milked two very fine house cows that became very lame with bad cases of foot rot, found that a tablespoon of copper sulfate in the evening bail feed cured it overnight. The cows were, of course, getting dolomite in their feed.

The disease causes smelly, suppurating and very sore feet, sometimes with large proud flesh growths forming in between the toes. If confronted with that condition, a sprinkling of straight copper sulfate on the growth after dipping the feet in the copper wash will help the proud flesh to disintegrate. The wash should be made up of two pounds of copper sulfate to two gallons of water and two pints of vinegar. The vinegar acts as a water softener to make the mixture soak into the lesions. Raising copper levels in the food, or giving the licks and maintaining the cattle at the correct level, is the quickest cure (and the best prevention), and there will be no recurrence even on the same land. However, if the farm has had artificial fertilizers used on it, the problem will be ongoing until the imbalances can be corrected.

Keratin, which depends on adequate sulfur and copper in the diet, is the component that gives skin and hair its strength. When foot rot (foot scald) starts, a thin, red line will be seen between the toes of the cow. This happens when the skin has inadequate keratin and is breaking down allowing the entry of the causative organism.

Foot Scald, *see Foot Rot*

Grass Tetany

This condition is caused by a deficiency of magnesium in the paddock, so it is not really a disease. Professor Ivan Caple, of the University of Melbourne, stated that "In dairy breeds, the risk of grass tetany is increased when potassium and nitrogen fertilizers are applied in autumn and early spring to promote pasture growth in late winter and spring." Organic methods of improving the pasture are far safer and more reliable in the long term.

Cattle will show signs much like lactation tetany which usually appears in spring with the rapid growth of grass, especially on paddocks that have had superphosphate applied and are, therefore, short of magnesium (and copper). The cattle will go down and die struggling, except in extreme cases where they die as if asleep. On that occasion, the vets called it superphosphate poisoning — not grass tetany.

The treatment is injections of magnesium and calcium which are available from the vet or fodder stores. The injections should be given as soon as the first signs are seen. If they're given in four places, each side of the neck and each side of the rump, they will act faster. On farms where bore water is used for drinking, the cattle on the bores do not seem to succumb, while those in a next door paddock on dam water will go down very rapidly. This is because most bores are high in magnesium.

This condition only strikes when the animals are on magnesium-deficient paddocks. A soil analysis of the farm is a fairly reliable guide to paddocks at risk for grass tetany. Once again, a good soil analysis can alert the farmer to many potential health problems on the farm.

Hardware Disease

This is the somewhat misleading name given to cattle with a depraved appetite which leads them to eat all kinds of inedible things including stones, bones and metal objects. If the object happens to be a piece of wire, it may pierce the gut and the cow dies. That is called "hardware disease" and the result is fatal.

Of course, the condition is caused by the deficient animal taking the law into its own hands and trying to meet its needs by eating anything that could contain the minerals that it is missing. If any of my goats lost their copper bells I would always find them in the dung of my land-lord's cattle, having been eaten and passed through.

Impaction (Constipation)

The manure becomes very hard, dry and difficult to pass. A drench of cooking oil should lubricate the works. Use at least a quart for a large cow.

Animals look hunched up and uncomfortable and should be drafted into a yard for observation and possibly a rectal examination. Impaction can occur on very dry food if the water supply is not good, or is extremely unpalatable, but it is a rare condition. Do *not* use liquid paraffin as it demineralizes the animal. Use an oil suitable for cooking, but do not recycle used cooking oil into the cows. Once oil has been heated, it is destabilized and can cause health problems in man and animals.

Infertility

The cause needs to be determined. If the cows do not come in season fully, the most likely cause is lack of copper. See that the cows have access to the lick in feed or *ad lib* or amend the diet if what they currently receive is insufficient. Cows that fail to hold to service (always assuming that the male is fertile) are, unless non-breeders, suffering from a lack of vitamin A. An injection of vitamins A and D, or A, D and E before the next heat will usually mean the failure will not occur again. Otherwise, supplementation with some sort of vitamin A coming up to service, or feeding the stock on a well grown green crop, would ensure they hold.

This sort of infertility is apt to occur after or during a long drought (which is probably how the native fauna are regulated). Particular care should be taken of the bulls, as in that case, vitamin A related infertility is usually irreversible in males.

A lack of selenium is another reason for poor or complete infertility in bulls. The sperm will be weak and few in quantity, and those that are there will tend to drop their tails. Luckily seaweed contains selenium in an organic form and making sure that stud animals receive their ration of the lick regularly will go a long way to ensuring sperm quality and quantity.

Injuries

For any injury where the skin is broken, tetanus must always be considered as a possibility. Tetanus usually takes 10 days to incubate, occasionally longer. Cleaning and disinfecting the wound really well helps to avoid the problem. Use either a copper wash, diluted peroxide or iodine. Giving either one tetanus antitoxin injection, or better still, daily injections of vitamin C for a two weeks or until the wound heals should prevent tetanus. Use 15 grams of vitamin C the first day and 10 grams thereafter. This can be administered intramuscularly or daily in the feed. See the section on tetanus in this chapter.

Body injuries

Cracked hips, shoulders, etc., are not necessarily the end of the road, especially if the animal will lie or stand quietly and rest the affected area. The sheer weight of big animals is the snag, and often the farmer has to be quite ingenious when getting a big cow back on its feet.

A very large Friesian cow belonging to a neighbor slipped into a drain and badly sprained or cracked her hip. The cow had a calf which fed easier when she was up so I helped him to lift it each day. We used the bucket on a big tractor and two girths that we made out of belting and leather. One sling I slid behind her front legs, attached the sling to the bucket and got him to lift her a few inches. Then I slid the other sling through just in front of the udder, let her down again, leveled up the two slings and then slowly got her to her feet. This was not as easy as it sounds, but we got better at it. This performance was regarded as sheer madness by all the other farmers in the district, but she was a valuable old cow and we thought it worth a try. Every day we got her up; she stayed up a bit longer each time with the calf happily having its meal. Finally she stayed up for several hours and then, about three weeks after the process started, she achieved it on her own. That cow was a particularly good foster mother, and the effort was well rewarded. The farmer will have to assess whether it is worth taking the trouble, I have always felt it was.

Occasionally after calving, a cow will not get up for a few days. Make her comfortable by propping her up with a few bales of straw or whatever. Give her good feed with all her minerals in it and water. Very often after a couple of days rest she will be back on her feet of her own volition. If the animal does not show any improvement within three weeks, it has probably sustained nerve damage from the calf being born, or when it was *in utero*, in which case the problem is irreversible.

Broken bones

The repair of broken bones will depend on the weight of the animal and the location of the break. A clean break through the front leg of a cow would usually be a shooting job. If repairable, healing should take place in 10-14 days provided the animal was not too old.

The cow will have to be immobilized and the weight taken off the affected limb (often the most difficult part). Bandage it into the correct position with a firm, but not too tight, bandage. Apply the splints, either flat pieces of wood or the metal type which looks like a ladder (obtainable from the vet), whichever works best, and secure them with a good strong bandage. This last bandage is best sewn into place so she cannot pull it off or catch it on snags. This should be left on until she goes sound and can put weight on the limb properly.

In the case of a compound fracture when the bone is sticking through the skin, thorough disinfecting of the wound is very important, followed by stitching, bandaging and splinting as before, but this is strictly a job for the vet.

Unless the animal is very valuable from a stud point of view, it is really more humane to put it down at once.

Skin wounds

For bad skin wounds, 10 grams of vitamin C for a few days will help healing along with about 10,000 units of vitamin E. The vets I worked with years ago in the United Kingdom taught me to disinfect thoroughly the first time. They said that disinfectants inhibited healing and should not be used more than once. Flint's Medicated Oil, if obtainable, discourages flies, but any good herbal ointment will help. Comfrey is excellent and so is calendula. If bandaging is necessary, try to make sure it does not stick; this is where the above-mentioned oils are so good.

When wounds are bandaged they quite often smell bad, but this is rarely a problem. Just clean them up and replace the dressing.

If there is a possibility of stitching a wound, it must be done straight away. It is no good calling a vet in the next day because once the skin dries, stitching cannot be done properly. Straight wounds can be very well stitched by the farmer. Use a strong curved upholstery needle (or have the vet leave you a supply of surgical needles for emergencies) and linen or similar thread. The thread, needle and wound must be disinfected. Sew the edges of the skin together with separate stitches and knot each one, leaving ends of at least one inch — any shorter and you may have difficulty locating the stitches when they need removing. It is really better to leave stitches exposed, as they tend to slough away if covered by a bandage. If flies are a problem, a little Flint's Medicated Oil smeared over the affected area daily should keep them away. After 10 days the stitches can be removed; swab the area with disinfectant, hold the stitch by one of the ends, and cut behind the knot. Pull out the stitch, wipe the area with Flint's Oil and leave it.

Tendon damage

A heifer with a seven-eighths severed tendon after it had walked through a mess of telephone wire on a road verge, was brought to me by a neighbor. The wound was on the hind leg. Stitching was out of the question as the tear was too deep and wide. I disinfected it thoroughly, then filled the wound with comfrey ointment — any antiseptic would have done, but comfrey is a great healer. I then covered the whole thing with a thick pad soaked in Flint's Medicated Oil and bandaged it up. The bandage was sewn on; I ran a flat splint down the front of the leg to take the pressure off the remains of the tendon by keeping the foot in position, and bandaged and sewed it on. We removed it one week later to see all was well, and then left it on until the bandage dropped off. The leg healed perfectly and the heifer never went lame. Often the wound did smell pretty terrible, but not gangrenous. Smelly wounds do not seem to matter with stock, keeping the bandage soaked with Flint's Oil will keep the wound supple and clean.

Johne's Disease (*Mycobacterium Paratuberculosis*)

This is a bacterial condition where the lining of the intestines become thickened and calloused and so the cow is unable to absorb nutrients from the feed for transfer into the blood stream as it should. The cattle die of slow starvation. There is also an enlargement in the lymph nodes noted at post-mortem, however this does not invariably mean Johne's disease; it can show up in several different conditions.

In my experience, there are two ways of contracting Johne's disease. The first is from the mother, where the disease lies dormant until the stress of calving brings it out. The other is from rank bad management. In both cases, the onset is made more rapid by mineral deficiency. The first method of transmission was long believed to be the only one.

I first read of Johne's disease in a book called *Goat Husbandry* written by a Scotsman, David Mackenzie. It is one of the few really good books available on goats. He dismissed all diseases with the observation that well looked after animals did not contract them. His goats had a huge spread of moorland and about three miles of coastline as well as being dairy fed, so it is not surprising that they enjoyed good health.

Mackenzie claimed that Johne's disease was a deficiency disease so I never really worried about it. A few years later, the top government vet arrived on my doorstep to test my herd for Johne's disease. I was not worried as I knew their feed was balanced. He was sure he would find positives because two goats I had sold to an aspiring goat keeper had gone down with the disease. As had the pairs of the other dairy breeds also bought from reputable studs. None of them had any sign of it either. That was when it first dawned on the veterinary profession here that Johne's disease could be contracted if farm conditions were bad enough, which they had been on the farm that had bought my goats.

Newman Turner, in his 1951 book *Fertility Farming*, describes how he bought a pedigree Jersey bull at a killer sale because it had Johne's disease and was condemned. The picture shows the bull as it arrived, a wreck, at four years of age. He grazed it on top-quality pasture with up to 65 different species of plants for a few months. The second photograph shows the bull at ten years, at which time it was winning prizes in the shows and had been doing its job for the last six years.

Acres U.S.A. reported a year or two back that animals getting the right amount of copper, cobalt, iodine and manganese do not get brucellosis and go into remission if they have it. This probably applies to most diseases, including Johne's disease, if we did but know.

My own experience with Johne's disease was some 12 years ago when I was milking goats in north central Victoria (I did not read Turner's work until 1996). I milked a herd of 35, half of them on lease. Unknown to me, a herd I took on had a high incidence of Johne's disease, which I did not find out until two years later. I sent home three unthrifty goats and continued to milk the remainder successfully. The Johne's disease positives had run with mine in one of the wettest winters on record.

Elaine, one of the leased milkers that remained, suddenly dropped in condition with frightening speed. It was a classic case of Johne's disease as I then found out when I got the vet to her. I never admit a disease is unbeatable until I've tried everything. In this case I gave her massive amounts of vitamin C injections (which is what doctors who know use for Crohn's disease). She was already getting her minerals. She recovered completely. But I later found out that she came from a long line of goats all of whom had developed Johne's disease and died in their turn. When the lease ran out, I asked for Elaine to have a full post-mortem when she died. The post-mortem showed absolutely no signs of the disease. She was the only member of her line not to die of it.

When the drought finally broke, hundreds of animals died right across northern Victoria as a result of sub-lethal nitrate poisoning due to capeweed (*Arctotheca calendula*), a South African importation. I lost 14 animals before I discovered the reason. The department of agriculture ran a post-mortem on the lot and kept informing me, in slightly bemused tones, that they did not have Johne's disease, which I knew. They were so busy looking for Johne's disease that they missed the real cause, nitrate poisoning, and it cost me my dairy herd.

I sold a six-month-old kid many years ago who was from a tough old milker (12 years) who had neither CAE nor Johne's disease (they were all tested then). I bought her back six years later and saw why the owners were willing to sell. They lived in the mountains and had the naive idea that a good milking animal needs no supplementary feeding or minerals which is not possible, at least not in Australia at any rate. Her tests showed Johne's disease, and my vet remarked I'd better do something about it as I'd cured Elaine years ago.

I segregated her, dosed her with two ml of VAM, two ml each of vitamins B1, B12, B15 and 17 ml of vitamin C (8.5 grams) all in a 25 ml syringe. I injected this dose daily for nine days. In four days she was back on the basic diet which, of course, included copper with *ad lib* seaweed, hay and her grazing. After that she never looked back. She was retested and came up clear. In her next lactation she got her milk qualifications having produced a very healthy pair of kids. It took ten months exactly before her manure was perfectly normally formed. I reckon the scarring took that time to heal and the intestines needed time to normalize which, interestingly enough, is the time that Newman Turner mentioned in his book. Certainly not an economic exercise unless it was, as in this case, a blood line that had no other representatives.

Johne's disease is certainly not economical to cure on a large scale, but it is perfectly economical to prevent. The two animals I saved were both very valuable bloodlines and high milking stock, as was Newman Turner's bull. The current protocol appears to be total eradication, as it is for foot and mouth, another disease that reportedly only strikes animals whose trace minerals are out of kilter.

It is definitely a problem in dairy cattle, especially intensively and conventionally farmed cattle. They will be low to nonexistent in their copper levels and Johne's disease likes a host low in that mineral. It is not spread by the pasture as believed; this could only happen if the cattle were too low in their copper. This is yet another important reason for knowing the mineral status of the paddocks.

Johne's disease is rare in beef cattle. I knew of one case where Angus (black) cattle were badly afflicted so the farmer changed to Herefords who managed to hold their own. This would point to the fact that a copper and iodine deficiency were the most likely minerals implicated in this disease as well as those mentioned below.

There seems little doubt that Johne's disease, like many other conditions, is related to bad husbandry and/or lack of the correct minerals. The organism is probably present in the soil of far more farms than is realized. Signs of the disease are wasting, diarrhea and recurring ill thrift — have the vet take a blood test. However, the only really sure diagnosis is on post-mortem since Johne's disease can show a false positive on a blood test.

Lactation Tetany

All tetany illnesses are basically due to a deficiency of magnesium. In lactation tetany, it is thought that calcium is also implicated. Although Hungerford said in 1951 that the old treatment of calcium alone works much better if magnesium is included as well, it is only recently that this seems to have been realized. Stock that has been receiving supplementary magnesium and calcium in the form of licks and are on chemical-free pastures should not succumb to lactation tetany.

The signs of lactation tetany are cows that appear uncoordinated, followed very quickly by collapse and then, if nothing is done, death. This tetany can occur any time after calving, usually fairly early on in the lactation. Cattle usually struggle in a circle while they are down until, after much struggling, the cow dies.

Basically, the milk production of the animal has used up all of its available calcium and magnesium, and there is not enough left to sustain life. The condition only occurs when the animals and/or the paddocks are deficient. Supplementation and paddocks that have been improved and top-dressed with the required minerals are the answer.

Treat the animal promptly with a calcium and magnesium injection (available from any fodder store) used according to instructions and the animal will soon be back on its feet. The vet who originally taught me how to treat this complaint pointed out that the remedy was far more efficacious if it was given in four doses, one in each side of the neck and one in each side of the rump as suggested for grass tetany.

LDA (Left-Displaced Abomasum)

According to a report in the *Weekly Times* (September 28, 1994), this is a rapidly increasing complaint among heavily grain-fed cows after calving. It is caused by over-feeding grain without the bulk feed to match. As pointed out in Chapter 6, carbohydrates in the form of chaff and hay, as well as the grazing, are *all* important for ruminants. LDA means that not long after calving the abomasum drops beneath the rumen due to the weight of grain. The gas in the rumen is then trapped and can only be released surgically, rather similar to bloat. Signs are poor appetite, decreased production with the flanks hollowing. Untreated cows die from a malady which is totally preventable.

Leptospirosis

This is a disease which strikes cattle whose mineral levels are not correct. In 30 years of keeping dairy goats who are, apparently, very susceptible to it according to the vets, I never immunized them nor did they once contract it; this in spite of dire warnings in years when it was particularly bad. Far more serious are the agitated calls I get from dairy farmers who have just completed their annual "lepto" immunizations with a death toll of three to four percent and an abortion rate in a similar ratio. This is on farms where leptospirosis had not occured.

There are two possible causes for these results, one is anaphylactic shock due to the cattle being already sensitized to the vaccine or, as in other immunizations, a sudden withdrawal of vitamin C from the tissues that is triggered by the vaccine. This is a well-known phenomenon. Standing by with the adrenalin might stop the former and heavy doses of vitamin C in advance, two tablespoons by mouth or 30 cc intramuscularly, might prevent the latter.

Lice and Exterior Parasites

Bad infestations of exterior parasites are caused by malnutrition. This does not necessarily mean that the animal is starving, but that the diet is unbalanced. Animals will be seen scratching and rubbing themselves against fences, trees and so on, and often the ears go bald. Lice are not very contagious in spite of what people may think. A healthy animal may have a few but they do not proliferate unless the beast is deficient, particularly in sulfur.

Modern chemical farming practices have made sulfur unavailable in farm produce, so all our feed is lacking in that necessary mineral. In 90 percent of the farm analyses that I see, sulfur levels are far too low (Neal Kinsey notes the same thing in the United States and worldwide). Feeding of sulfur in the licks will help to keep animals free of lice and other exterior parasites. As long as the sulfur does not exceed two percent of the diet it is quite safe. This means that a cow could be given a heaped tablespoon a day if she had an infestation; the lice would gradually leave her over a period of five or six days.

When cattle become heated for any reason the presence of lice usually shows up, as they leave the skin and crawl out on the hair. Exterior dressings of sulfur are fairly effective, just rubbing in two or three handfuls of sulfur along the spine is enough.

As pointed out in the section on sulfur, a lack of this mineral means that cattle do not absorb and digest their feed as well as they should. Bringing up the sulfur levels in the soil by top-dressing with gypsum (after an analysis) is the long-term remedy.

Liver Fluke, *see Worms*

Mastitis

Mastitis is quite easily controlled by ensuring that all the animals get the right amount of minerals in either hand-fed or on-demand licks. Bail-fed milkers will need a three tablespoon (30 grams) per head per feed minimum. Generally, feed is grown with artificial fertilizers and is low in the necessary lime minerals and copper. The protein levels in Chapter 6 should be adhered to since too much protein in the food is a potent cause of mastitis and overly rich diets should be adjusted.

If the pH of the ration is correct and the lactating animal is receiving its lick ration, the type of the mastitis organism seems to be immaterial. Even a torn udder will not produce instant mastitis. Naturally, in this case, the farmer will treat the tear and give the animal extra vitamin C and dolomite to prevent any infection.

An infected cow should be given an extra tablespoon of dolomite and the same of vitamin C night and morning until the infection clears — usually three to five days. In the United Kingdom a teaspoon (five grams) of copper sulfate has to be added to that mixture as the basic pasture is higher in protein than in Australia.

We now have a new or ancient modality, according to how one looks at it. Hydrogen peroxide; 10 ml squirted straight into the affected quarter has cured black mastitis in hours. We used to take ten days curing it with massive amounts of vitamin C.

Additional vitamin C could be given by mouth for a day or two as well. The advantage of all the above methods is that the quarters are not lost as is so often the case when ordinary drugs are used.

Metritis

The signs are usually an evil-smelling discharge from the vulva following calving, often after the calf has been born dead or taken away in pieces. If a vet is attending, he will insert a pessary, otherwise give a washout of a teaspoon of salt to liter of water to remove the worst of the infective material. If an animal sheds the afterbirth cleanly, this should not be necessary.

Metritis and other conditions affecting the uterus are mainly caused by a lack of vitamin A. In a bad year with much dry weather there are often quite a few affected animals, as they will not have had enough green grass to obtain the necessary vitamin A. There will be a predisposition to metritis in stock on chemically manured paddocks as the chemicals interfere with the synthesis of vitamin A, and the cattle do not receive as much as they should.

Any animal with metritis should be put on a course of vitamin A in some form or other. Vitamin A, D and E injections are suitable as prescribed on the bottle, or an A and D (cod liver oil) drench, 20 ml twice a week if possible, will work. Cattle usually take the oral dose quite well on feed. Vitamin C will also help clear up the infection; use either 7.5 to 10 grams by injection every other day for a cow or 12 grams orally every day in the feed for a week.

Milk Fever

Milk fever occurs when the cow's calcium reserves are too low for it to sustain life. This occurs shortly after calving. As with lactation tetany, the animal has put all it has into the milk and left too few minerals for herself. The signs are exactly like snakebite, curiously enough, lethargy, slow movements and the pupil of the eye appears much dilated as the eye muscle relaxes. Death will follow if treatment is not started immediately (in either case). The magnesium/calcium injection for the relief of milk fever is obtainable from any feed store and the amounts are given on the bottle. For best results, put the injection in each side of the shoulder and rump, four places in all. This way it is dissipated as fast as possible.

Stock that has been fed on properly balanced paddocks and/or been receiving the basic lick, will not be prone to milk fever. I have found that even confirmed milk fever sufferers will not have a recurrence as long the necessary minerals have been provided. (Often if a cow has had milk fever once, she seems to be more likely to contract it again.)

Bearing in mind that snakebite and milk fever present exactly similar signs, this alternative should always be considered. It sounds unlikely, but if there is any doubt treat as for snakebite as well; it will do no harm. (In the middle of a cold winter I was once confronted by a horse in an unlikely advanced state of pneumonia and suffering from snakebite, luckily the same treatment worked for both.)

Navel Ill (Polyarthritis), *see Arthritis, Infective*

Nasal Bots

This is a small insect that looks like a beetle larva without wings. It is about half an inch in length when fully grown. Sometimes one will hear cattle sneezing and if they happen to be on concrete, one may find a bot that has been sneezed out on the ground. The egg is laid by the adult fly in the nostrils and animals can be heard snorting when the flies are around. If the stock can be handled, putting Vicks or K7 on the nose will often discourage the flies or make the beast sneeze hard enough to dislodge the larva as it crawls up into the nasal passages.

This can be serious if the cattle are very young, as the nasal passages may not be large enough to accommodate the larva, so it migrates into the brain or elsewhere in the head. Fortunately, in full-grown cattle there is enough room in the nasal passages for the bot to cause nothing more than discomfort. But in the brain it can cause signs similar to circling disease in sheep, due to brain damage, in which case it is generally fatal.

Pink Eye (Conjunctivitis, Ophthalmia, Sandy Blight)

Runny eyes are often the first sign of this illness, then they cloud over and look opaque. If treatment is not started promptly the eyeballs swell, ulcerate and burst — very painful and apt to cause permanent blindness.

Pink eye is caused by an organism that only operates if the host is deficient in vitamin A. It is highly contagious, but will only be caught by other animals deficient in that vitamin. In Australia, where huge areas are dry and without green feed for long periods, this can be a problem. It is made worse by the use of artificial fertilizers which inhibit all vitamins to a degree.

Vitamin A is stored in the liver and there should be, in theory, enough from the wet season to see a beast through the dry, but prolonged drought and poor land may cause problems.

Geoff Wallace, the inventor of the Wallace Soil Conditioner, had a mob of Texas Longhorn bullocks that contracted pinkeye on a poor paddock. Half of his farm was already converted to organic methods, so he moved the bullocks onto a healthy paddock and the pinkeye cleared up in a few days. Easier than manhandling the beasts.

To treat pinkeye the sufferers must be yarded as soon as possible. The affected eyes can be treated by pulling up 20 ml of cod liver oil and squirting three ml into each eye and the rest (14 ml) down the throat. This may be repeated for a few days if necessary. I am indebted to my local vet, Alan Clark, for that remedy and it certainly works.

Plant Poisoning

It will always be possible to find plants that are potentially poisonous in any pasture. The nightshade family is particularly widespread, but unless animals are starving, they will only nibble at undesirable plants and take no harm thereby.

Some plants and trees can be tolerated at times and appear to be quite dangerous at others, again the amount eaten must have some bearing. In the old days in the United Kingdom, goats had a reputation for preventing "contagious" abortion in cattle, so a wether was kept with the herd. This was not strictly true, because the abortion was caused by a poisonous weed not an organism. Goats happened to like the weed and ate it; it did not affect them and so they stopped the cattle from becoming ill.

In my experience, the following plants can poison animals occasionally.

Boobyalla trees

My goats, and occasionally horses and cattle, ate from them freely with no ill effects at all. Other people told me their stock had found them very dangerous.

Bracken

This plant was never considered a problem in the United Kingdom; in fact, its herbal qualities are considered valuable there. Here in Australia it is a cumulative poison and causes bone marrow damage. I suspect that poisoning arises because animals are shut in paddocks with nothing but bracken to eat. If this happens several years in succession, inevitably the poison builds up to dangerous levels. When there is good feed available, stock hardly touch bracken. Bracken only grows on poor, potassium-deficient pastures and top-dressing with animal manures, soil aeration and improving the health of the paddock very soon discourages it.

Capeweed (Arctotheca calendula)

This scourge, which came from South Africa, is sometimes the only source of feed, particularly after drought where there are no competing species. It also takes over on very poor land after excessive wet, when the paddocks are badly pugged and again there is no competition from grasses. It likes a low pH and a compacted, anaerobic soil.

Capeweed can be excellent fodder and animals milk well on it, *provided* the farmer remembers two things. It causes a very bad depletion of magnesium and supplementary dolomite will have to be fed while capeweed is the chief source of food. If the stock are already on dolomite, they will probably need to have their ration doubled. Animals on capeweed soon start to scour and the normal remedies do not stop it until extra dolomite is given. As soon as the capeweed ceases to be the main source of food, the extra dolomite should be stopped as too much may inhibit copper uptake.

Secondly, capeweed depletes iodine to a fatal degree. This is more insidious as often the cattle do not start to die until after the capeweed has died back. The better the cattle look, the quicker they seem to die. Both these effects are caused by the high nitrates in the capeweed. However, if the extra dolomite and iodine are fed, cattle seem to be able to tolerate the extra nitrates. The iodine could be fed as liquid seaweed in the water, seaweed meal, or as Lugol's Solution, obtainable from a vet, who will advise on the dosage. Capeweed hay does not appear to have the same effect as when it is fresh; it is difficult to dry, but good feed when it is harvested successfully.

Capeweed is at its most dangerous in extreme drought or very wet conditions where there is little sunlight. Sun- light triggers an enzyme called nitrate reductase which reduces the nitrates to amino acids and proteins which are digestible. When this does not happen, the nitrates turn to nitrites in the rumen and poisoning ensues. In this type of poisoning, the blood shows up almost black due to lack of oxygen.

Professor Selwyn Everist, whose work, *Poisonous Plants in Australia*, covers nitrate poisoning, states that it is unwise to use MCP or 2,4,5-T type sprays to kill capeweed. This practice enhances the dangerous effects and makes the plant even more palatable to stock.

This advice applies to any broad-leaved species wherever they are. When I lost my herd of milkers from this ailment, I was not feeding the stock lick as such; had I been, I think the outcome would have been different.

Eucalyptus

Only the very young fresh shoots cause trouble as a rule. Consult the prussic acid paragraph in the next section on poisons.

Heliotrope

A small, silvery-colored, grey-green plant with clusters of pale mauve flowers. It is very high in copper (150 ppm) and causes the "yellows" in stock, in other words, jaundice from liver damage. Adequate dolomite in the diet stops this effect and it should be available in licks or bail feed. As a rule, the jaundice does not occur until the heliotrope has died off and the cattle start to graze green grass again, so it is very important to keep the dolomite up after the heliotrope has died off as well.

Top-dressing with the required minerals and adjusting the pH discourages this weed almost completely.

Lantana

This is a shrub with widely differing reports. One says the red one is poison and the yellow is not, another says it is a good source of feed, and yet another says it is lethal. Perhaps best avoided.

Lilac

This can be poisonous to calves. In grown cows it is tolerated, but can and does poison the milk. In the drinker, the symptoms are vomiting, diarrhea and discomfort. Children suffer more severely from the above conditions, which cease when the source of the trouble is removed.

Lobelia

This is described in Mrs. Grieve's *A Modern Herbal* as containing an alkaloid that is a strong poison. It is known to affect and kill cattle and sheep on occasion. Dolomite and vitamin C orally, a tablespoon of each every half hour, as well as injections of vitamin C intramuscularly could possibly work.

Patterson's Curse (Salvation Jane)

This is a bright purple flowering plant with big leaves which, in the spring, carpets whole areas of deficient land with a low pH. It is similar to heliotrope and St. John's Wort in it's action. It is a relatively deep-rooted plant and does, therefore, grow on land which is deficient in copper on the surface. Treat the land as for heliotrope.

Peach and plum trees

These are safe when fresh, but cyanide forms in the leaves when wilted. Dolomite and vitamin C could be tried if there is enough time. Keep cattle away from them.

Privet

Same as lilac.

Rhubarb

The leaves of this plant contain oxalic acid. Oxalate depresses calcium and iodine. My goats used to make a beeline for rhubarb leaves whenever they got in the garden, with absolutely no ill effects. Perhaps the dolomite in their diets protected them. Certainly dolomite or ground limestone can be used as an antidote.

St. John's Wort (Hypericum)

This is a plant which grows about a foot or so high, with thin leaves and yellow flowers. It is another plant about which there are conflicting reports. Basically it is very high in copper, high enough to cause trouble in white animals, but black ones tolerate it and indeed seem to thrive on it. Dolomite could be tried as an antidote as it does neutralize copper in an overdose. Like Patterson's Curse it can, and often does, grow on land that has a surface deficient in copper and, like Patterson's Curse and Heliotrope, it does not like well-balanced, healthy soils.

Following are plant poisons that are very dangerous:

Azaleas

The antidote to these is oral vitamin C, and for a cow two tablespoons (40 grams) made into a drench should be enough to effect recovery. Stock collapse and become moribund very quickly from this poison.

Black Nightshade (Solanum nigrum)

This is a plant of the nightshade family. It does not appear to be very poisonous unless starved animals are limited to eating it alone. In my experience animals leave it alone anyway.

Deadly Nightshade

Belladonna is the really poisonous type and is found in Europe and possibly the United States. Try vitamin C to regain health.

Oleander

This plant is deadly. Possibly vitamin C would work, but has not been tried. I once saw a goat happily eating an oleander on a nature strip, and judging by what was left of the other bushes, she had been doing it for quite a while, but I would not recommend trying it. Even a scratch from the wood is deemed very poisonous.

Potatoes

These are poisonous when green or as potato haulm. These contain solanine, a highly dangerous cumulative poison. There is probably no antidote. Vitamin C could be tried at a dosage of two tablespoons (40 grams) for a cow.

Yew tree

Deadly at certain times of the year, and the trouble is no one knows which ones. I did not know this years ago as I waited for a horse to die after I found it had got into the garden and was eating yew. It lived.

I would regard most garden shrubs as suspect and refrain from feeding them; many blue flowered species are poisonous. There are plenty of good fodder trees, coprosma (mirror bush), apple, pear, and nut trees, most acacias, kurrajongs, lucerne (tagasaste), casuarinas and paulownia to name a few. All are safe to feed if necessary.

The following plants are sometimes safe.

Linseed (rape, flax)

This contains prussic acid and should be used as a strip grazing crop. Allow stock on it for an hour or two each day; more than this could cause problems. Linseed grain, if heated, must be boiled for four hours to destroy the prussic acid effectively. In this form, it is an excellent feed additive for putting on condition and milk (see Chapter 6).

Phalaris

This grass (and rye grass) when grown on artificially fertilized soil, especially soils that are sick, will cause trouble — usually staggers. It is quite difficult to treat, but vitamin B1 could be tried. Both grasses, grown on remineralized fields, are excellent, high-quality feed, either fresh or as hay. Of course, the more diversity that there is in a grass ley, the healthier the animals will be. I find that one of the joys of getting the lime minerals in balance is that many varieties of beneficial plants reappear.

Rye grass

Same as for Phalaris above. It is an excellent fodder on healthy paddocks.

Sorghum and Sudex

Both contain prussic acid when young, and should not be grazed until they are over one foot high.

Pneumonia

Signs of pneumonia are labored and rapid breathing. If the ear is laid to the chest, it sounds rather like an express train in a tunnel. A high temperature, misery and occasionally coughing will also be noticed. Ordinary pneumonia can occur in any animal that is below par and subject to temperature stress allied to, in nearly all cases, poor nutrition in which the calcium and possibly magnesium are too low. Bad housing with too little air is also a frequent cause of pneumonia in young animals; it is not a good idea to keep calves confined. If they have to be kept in sheds, they should be airy, but not drafty.

This kind of pneumonia is generally bacterial and will respond to good nursing, massive doses of vitamin C and vitamins A and E especially. Twenty grams of vitamin C by intramuscular injection every two hours, with 10 cc of vitamin B12 daily (given with one of the vitamin C injections) and vitamins A, D and E either orally or by injection should help. Some firms market injectable vitamin E. White-E is a powder that can be used orally in food (the dosage is on the container). Vitamin E is invaluable for the convalescing pneumonia patient because it helps heal lung damage. The extent of lung damge can be gauged by checking the breathing rate and the vitamin E may be given for a week or two until the breathing rate improves.

Pleuropneumonia

The dreaded "Pleuro" (mycoplasma pneumonia) is another problem altogether. Although it can affect all species, in most people's minds it is *the* dread disease that decimated the cattle numbers in the early days in Australia (respiratory problems are always worst where calcium and magnesium are low). It is highly contagious and for this reason, any case of pneumonia should be checked out by a vet.

Mycoplasma pneumonia has had a habit of recurring once the animal has contracted it and it seemed impossible to effect a permanent cure. However, that was before some of the drugs we have today were obtainable. Erythromycin, in particular, is highly effective against mycoplasma infections. The treatment, as above, for ordinary pneumonia should be given at the same time as any drugs are given. Needless to say, any animal with pneumonia should be segregated from the healthy ones immediately, pending the arrival of the vet.

Poisons

General

Poisoning due to chemicals is extremely difficult to cure, if not impossible in some cases. Large quantities of vitamins A, B, E, C and zinc are suggested for humans suffering from inorganic sprays or similar type poisonings. These could be tried if the animal showed some hope of living, and they have occasionally been successful.

Sometimes removing the cause is effective. Three goats from three different localities developed leukemia that showed up as edema from the chin to the top of the forelegs. All of them had been in contact with pesticide sprays in various forms. Taking the animals from the source of contamination and feeding them good, healthy grazing on an organic pasture, resulted in a complete remission in a couple of weeks.

In all cases of suspect poisoning (except 1080, see entry to follow), it is always worth trying doses of vitamin C in fairly large quantities, both orally and by injection. Vitamin E is the other really useful vitamin for poisons because its healing action is very helpful, particularly when used in conjunction with vitamin C.

Arsenic

Arsenic poisoning produces a rather sweet smell in the mouth, extreme distress, vomiting (even in a ruminant) and a deathly chill. I have been told that ruminants die if they vomit, but I have managed to bring an animal through by using large doses of vitamin C, both by injection and orally, along with vitamins E and B12. Sadly, the exercise is not very productive as excess arsenic causes hormone and bone marrow damage and the animal I saved was never very good again.

Lead

Lead poisoning is often caused by an animal licking paint or old paint cans and results in anemia, muscle weakness and diarrhea. Drenching with Epsom salts helps clear the system, and the usual supportive measures should be followed. In the past, this was quite a common form of poisoning in young calves. When they are missing nutrients, like their parents with hardware disease, they will try anything in an often vain attempt to correct the deficiency.

Mercury

Hopefully, mercury poisoning is becoming a thing of the past now that it is banned as a silo fumigant. In its severe form, it causes shredding of intestinal mucosa and diarrhea. In the sub-acute form, often caused by mercury-based latex paints, digestive disturbances, loose teeth and salivation can be the warning signs. In either case, call a vet immediately and remove, if possible, the source of contamination. As in other cases of poisoning, supportive treatment with vitamins C and E, will be of help.

Nitrate

This poisoning is mostly covered in the section on capeweed above. Nitrates are stored in broad-leaved plants, even clover on occasion, and, in certain cases, turn to deadly nitrites in the gut. Normally nitrates are processed to proteins and amino acids in the plants by an enzyme called nitrate reductase. This process needs sunlight. In conditions of extreme drought, when there is a lot of cloud cover, the nitrates build up to dangerous levels turning to nitrites which deplete the oxygen in the blood of the animal that eats it, and the animal dies. On post-mortem examination the blood appears completely black. At first, the obvious symptoms are almost identical to tetanus. In fact, my vet and I took 24 hours to realize that it was not tetanus when we saw our first case. According to Everist, in *Poisonous Plants in Australia*, huge doses of vitamin C are the only hope, but we did not find it worked. Sweet-smelling dung is one symptom of nitrate poisoning.

Phosphorus

After eating phosphorus-based rat baits, the animal has a craving to drink and must not be allowed water in any form, as this causes the phosphorus to react with the liquid and burn the intestines. If a phosphorus-poisoned animal has had a drink, it is kinder to put it out of its misery.

Egg white, glucose and possibly a very little milk may be given at hourly intervals until the beast shows signs of recovery. This worked with a dog I cared for and it took 36 hours. It would be very difficult to do with a much larger animal. In this type of poisoning the breath of the animal also has a sickly sweet smell, quite unmistakable, and the sufferer is obviously feeling heat in its gut.

Prussic Acid

Poisoning from prussic acid usually comes from young sugar gum leaves or other young eucalyptus. The animal will foam at the mouth and go down. Drenching straight away with pharmaceutical chalk or fine dolomite will effect an immediate recovery by neutralizing the poison. Give at least four tablespoons of chalk or dolomite to a half liter of water for full-grown cattle.

Spirodesmin

This is also called facial eczema and is found in the pithomycetes spore. It causes fatal toxicity. It has caused death in alpacas and could affect cattle. High doses of vitamin B1 as recommended for thiaminase poisoning, vitamin K and supplementary zinc in the form of zinc sulfate could be effective. Consult your vet.

This toxin only seems to occur on sick, acid paddocks with a low pH and poor mineral levels. It would not happen in paddocks that have been tested and top-dressed with the appropriate lime minerals, i.e., gypsum, lime or dolomite. Hot, humid weather will show up these kinds of organisms after cutting very long grass in warm weather, when the mold spores start to grow. Slashing should be done frequently so there will not be heavy, wet grass lying on the paddock which will encourage this condition.

Thiaminase

This poison works by destroying thiamin or vitamin B1. It is present in molds and is one of several poisons that is fungal in origin. See the section on vitamin B1.

Urea

Urea poisoning can be a problem whenever urea is fed to cattle and the farmer should try to keep them under reasonable observation. Signs are stomach pain, uncoordinated gait, muscle tremors, deep, slow, labored breathing, weakness and collapse, bloating, frothing at the mouth, vomiting and violent struggling at death.

The only treatment I have heard experts recommend is four to eight liters of equal parts water and vinegar. I would suggest adding a two tablespoons of vitamin C powder to that mixture.

Cattlemen have reported that when feeding the basic lick to beasts being fed trash and low quality feeds, urea does not need to be fed. Perhaps the sulfur in the lick helps.

1080

This is also called fluoroacetate or 23 ppm fluoride. Glycerol monoacetate is the antidote, which is not normally carried by the veterinary profession. If 1080 poison is known to have been taken by an animal, the antidote must be given within 20 minutes. It will take about four hours for the beast to die after that, in great pain, and nothing can reverse it. Shooting is the only humane answer unless the antidote can be given in time. I am told by someone who worked in the Lands Department that 1080 does not cause pain. The vet to whom I took a neighbor's dog dying from it will testify that it certainly does in dogs.

Pregnancy Toxemia

This is a fancy name (so is twin lamb disease) for a condition that occurs when the fetus has taken all the necessary minerals and the mother is left without enough to sustain life. It should not arise in well-kept stock, even with multiples *in utero*, which is very rare in cattle.

The signs are increasing lethargy. The pregnant animal makes a lot of fuss about moving or rising after rest and finally refuses to do so at all. Exercise undoubtedly helps, as it dissipates the minerals that are available around the system and keeps the gut working, but the time inevitably arrives when the minerals run out. Then mother and young will soon die unless swift action is taken.

The quickest and most effective measure is to give a large drench, 50 ml, of seaweed extract to which a dessertspoon (per cow) of the basic lick has been added. This usually gets them on their feet very quickly. The dose should be repeated every six hours until the cow is up and then half the dose may be given daily for a few days. A system of feeding should be started at once giving her all the minerals she needs right up to calving, and do not forget the cider vinegar. Many farmers often do not bother to give extra feed until a cow is actually lactating.

Supportive injections of vitamin C, with some added vitamin B12, will also help if the patient is very low. In the old days glycerine was used for this condition and it did keep the cattle alive, but in no way relieved the problem and animals seemed to hate the taste of it.

Prolapse

Prolapse usually occurs before calving and part of the placenta will obtrude from the vulva; this is generally more noticeable if the animal is lying down. Prolapse is due to lack of muscle tone and can be cured by giving the cow calcium fluoride tablets over a period of two or three days. (Calcium fluoride is nothing to do with the sodium fluoride in the water supply.) The tablets are obtainable at any health shop in the homeopathic tissue salts area. The dosage for a cow would be 50 tablets crushed up and given by drench or in the feed once a day. They are quite palatable. It usually takes something less than 48 hours to bring about a cure.

Do not on any account cut the prolapse, unless you want a dead calf and cow. It should be pushed back gently, but will generally not stay in while the muscle tone is too weak to hold it there.

The device usually suggested to hold in a prolapse has a very poor running record. It often punctures the placenta before birth, resulting in the loss of the amniotic fluid (needed for lubrication as well as protection), and a dry birth usually means the loss of calf and cow. Try the calcium fluoride, it has worked several times with afflicted stock. This complaint does not occur on remineralized fields or in stock that is getting the basic lick.

Prolapse after birth

The old folk practice used very successfully in the James Herriot books was putting the tissue back with a couple of kilograms (about five pounds) of sugar. This was used for a cow when no other remedy was successful. The calcium fluoride would also be of help. But I have found that so many of these odd ailments are no longer a problem when animals are correctly fed.

Basically the animal is deficient and if the trouble is confined to one beast, unless she is very old, one has to assume she is not thrifty and cull her.

Red-water Disease (Tick Fever)

Vitamin C could be tried for affected cattle, it detoxifies tick and other bites quite effectively. Thirty ml (15 grams) of vitamin C by injection every two hours is the recommended dose. However, prevention is much better than cure in this case as in others. Raise the sulfur levels in the animals by putting out the basic lick. If the soil is out of balance and the outbreak continues, the sulfur could be raised to 10 pounds for a while. It is unwise to let cattle have molasses in any form as it encourages biting insects.

Retained Afterbirth

This should not occur in any cows unless they are low in potassium, selenium and possibly vitamin A. That said, the septic effects can be offset by injections of large doses of vitamin C. I had a goat that failed to pass an afterbirth due to trying unsuccessfully to produce two kids simultaneously in the middle of the night. By the time I extracted the dead kids, there was no chance of getting out the afterbirth without being very rough or resorting to drugs. I gave her 10 grams of vitamin C by injection in the muscle three times during the day and again next day. About a week later she was slightly off color so I repeated the treatment for two days, after which there was no trouble and she produced quite normally next time around. Other people have reported the same results to me. A cow would need three times that dose at least (30 grams). The afterbirth is obviously reabsorbed *in utero* in such cases.

As in dystocia, giving cider vinegar coming up to calving, along with seaweed products to supply the selenium, would be preventative measures worth taking. Vitamins A and D should always be given, especially in dry years.

Rickets

This is a disease that can affect young animals that are not getting the right amounts of calcium and magnesium in their diet. Occasionally it can be due to the fact that the calcium and magnesium levels are good, but there is a lack of vitamins A and D and/or boron, without which the calcium and magnesium cannot be used correctly.

A sign of rickets is generally legs which bend under the weight of the body. Children from the slums in the last century often developed rickets to the point where they had to wear leg braces, mostly occasioned by the lack of sunlight and no milk in their diets.

Very occasionally, diets too high in phosphates can have the same effect by upsetting the phosphate to calcium/magnesium ratio. Adjusting the diet accordingly will bring about an improvement. Make sure the calves get some of the lick as soon as they can take it in their feed. Do not over feed milk.

Ringworm

Ringworm is a skin disease caused by a fungus, not a worm. The name originated because the fungus works in concentric rings and looks like a coiled worm. A wash of copper sulfate, two tablespoons in a pint of cider vinegar, rubbed well into the lesion usually effects a cure straight away. If not, repeat the application until it does. Ringworm is not contagious except when animals are deficient in copper. Cider vinegar rubbed in well will also effect a cure and might be used if the ringworm is too near an eye to use the copper.

Making sure that the cattle's copper requirements were met would mean they did not catch ringworm.

Ross River Fever

This is a debilitating disease carried by biting flies. The beast is ill and the joints are swollen and sore — just like afflicted humans, in fact. The sulfur in the lick should stop the flies. Do not feed molasses on any account as it makes the flies bite worse. Large doses of vitamin C will bring it under control along with the right minerals. The addition of a dessertspoon of powdered ginger daily for a few days, then twice a week, should finish the cure. The latter is an old Chinese remedy.

Scabby Mouth (Orf)

A herpes-based infection, it is a particularly fast spreading and debilitating disease where the affected cattle become too sore to eat. It starts as a small scab by the mouth and quickly spreads all over the lower part of the face and it is *not* foot and mouth disease. The same wash as for ringworm works very quickly. Dip the affected animal's face in the mixture if possible, otherwise one or two applications of the mixture rubbed well in will cause the scabs to clear up. The disease does not have to run the mandatory three weeks. It is highly contagious if animals are copper deficient. Amend their diet straightaway.

Scouring, *see Diarrhea*

Snakebite

In some districts snakebite is the cause of significant losses among cattle, especially calves, who are more curious than adults and wander more. Large animals do not, as a rule, appear to take any notice of snakes one way or another. In Australia, Tiger snakes seem to be the worst offenders, being rather resentful of intrusion. Although highly venomous, their biting mechanism is poor; however, they seem to attack stock more often than other varieties (they are similar to English vipers or adders). Brown and black snakes tend to get out of the way.

Snakebite either kills immediately or it takes an hour or more according to the locality of the bite (and possibly the mobility of the patient). Signs of snakebite are an uncoordinated beast, collapse and very enlarged pupils. In some cases, if the bite is near the eye, one or both will cloud over too. Generally the bite cannot be seen. It becomes apparent about three days later, when the hair dies away around it and if squeezing out the venom, take care.

We are indebted to Dr. Klenner in the United States for discovering that vitamin C was a complete snakebite cure. The type of snake is quite immaterial because the ascorbate detoxifies the venom, whatever the type. Antivenom, apart from being expensive and that you need to know the type of snakebite, can also cause anaphylactic shock. I'm told this is rare, but I have seen it twice after antivenom was used. Then, if no adrenalin is handy, it may become a case of the shock killing the cow and not the snake.

In snake areas the farmer should keep two or three bottles of injectable vitamin C (100 ml bottles — two ml to a gram) in the fridge or cool room. Cattle need 25 cc (12.5 grams) in each side of the neck. A stout needle will be needed, size 18 or larger. This should be followed up in a hour or so by the same amount again. The vitamin C can be given orally by drench, in which case the equivalent of at least 30 grams (six teaspoons of powder) should be given to a cow. There is no danger of an overdose; it merely wastes money. Calves would need at least 10 cc by injection. Continue to give the vitamin C to the patient until it looks really well again; this usually takes one day.

One area where snakebite does not respond to the above suggestions is the udder. The cow will not die if treated, but the udder will be ruined. One method I have not tried could be to find, if possible, where the bite is and inject a large amount of vitamin C (100 ml) straight into the udder (through the wall). It would be worth trying since nothing else works. Ten mls of 6% H_2O_2 in 12 ml of rainwater works even faster.

Running a herd of commercial milking goats in a very bad tiger snake district, I resorted to putting bells on them. This cut the snakebites to nil. I know snakes are not supposed to hear, but they must have picked up the vibrations. Cow bells were used in the early days and are still obtainable.

Spider Bites

Spider bites are usually not so serious as snakes and ticks. Southern spiders (red backs, etc.) seem to cause swelling at the site of the bite and vitamin C helps these clear up. The swelling usually takes two or three days to go down. The swelling is only serious if it is in the throat area and quick administration of ascorbate usually stops it becoming any worse. Otherwise treat spider bites exactly like snakebites.

Note: in *all* poison bites, *if* the site of the bite can be found, rub sodium ascorbate well into it after treatment with the injections, etc., as it stops the pain.

Tapeworm, *see Worms*

Tetanus

The old name for tetanus was lockjaw because, shortly after the onset of the illness, the jaw is locked. Animals appear to stagger and seem disoriented. If hit smartly under the chin, the eyes will roll backwards.

Tetanus is caused by *Clostridium tetanae*, a clostridial bacteria that proliferates in deep airless wounds that have not been disinfected properly (often because they are not even visible). Shotgun pellets are very prone to setting up tetanus. It takes about 10 days after the wounding before it appears. Piercing wounds that the farmer often does not see and those from rusty nails are particularly dangerous.

The tetanus organism is generally present in soil where stock of any kind have been running. It used to be associated with horses, but it can occur with any animal, so if seen, all wounds should be thoroughly disinfected —

iodine, alcohol or disinfectant may be used. Vitamin C must be given after tetanus prone wounds until all danger is passed — at least 14 days — unless the vet has given a tetanus injection.

Tetanus is a horrible disease which takes several weeks to run its course. The animal is in a nervous and painful state with a high fever and the slightest noise causes it to go into convulsions — very difficult in a large heavy cow. Top class nursing was the only hope prior to the use of vitamin C. Fortunately it works very quickly with tetanus; generally three hours or less and the patient is usually well on the way to recovery.

Vitamin C should be given by injection and, if the jaw is locked, it is difficult to do anything else. For a large animal a whole bottle, 100 cc (50 grams) should be given initially, and 30 cc every hour thereafter until the signs cease. When the symptoms do subside, the battle is won. It is not good waiting several days and then expecting the vitamin C to work; the quicker the treatment is started the better.

After improvement, the animal must be kept very quiet and the vitamin C injections cut back to twice a day for a couple of days. Once the beast is up and about, just keep an eye on it. Light, nourishing food and its minerals should complete the cure.

However, if the cause of the tetanus has not been determined, examine the animal very carefully, because when the actual tetanus is obvious, the wound will be very infected. In the case of a bullet wound, or some other imbedded metal, it may be necessary to extract the object. I saw this once in a horse that had been hit in the leg by shotgun slugs.

Tick Bite

Vitamin C works well for tick bites, even in cases where the animal is already in a coma. The vitamin should be given as for snakebite. If the beast is in a coma, take care if giving it orally that it does not go down the windpipe. Find the tick(s) and remove them carefully. Cattle on the lick should not get attacked by ticks; these are only a nuisance on low grade, out of balance fields.

Tick Fever, *see Red-water Disease*

Three Day Sickness (Bovine Ephemeral Fever)

This is an illness carried by mosquitos. It causes a high fever that reputedly lasts three days. Farmers tell me it can last three weeks or more. Megadoses of vitamin C could be tried, 20 cc every few hours for a large animal, continued until it improves. However, prevention is the best approach and adding sulfur to the feed, as in licks, is the answer. When I suggested this to a farmer who rang, he suddenly realized his cattle had not contracted it the year he gave them a sulfur lick. The sulfur in the standard lick may have to be doubled if the outbreak is bad. Those that recover should be immune to further attacks.

Toxoplasmosis

This illness causes cattle to abort or bear dead calves. It is carried by cats, and really quite impossible to stop because cats cannot be controlled. Like the above ailment, lifetime immunity usually results from an infection. It is always wise to have a vet check out a dead calf to see if toxoplasmosis is the cause. The disease is a zoonose, and particularly can affect pregnant women who should be *extremely* careful about handling dead or aborted calves. They should also use caution when handling cats and should not handle litter boxes under any circumstances. This is another illness that we have not seen since giving the high amount of copper.

Tuberculosis

When we first came to Australia in the late 1950s, I was somewhat startled to find that tuberculosis was still a fairly common ailment here. Even these days there are sporadic outbreaks periodically. The same minerals that, when absent, allow brucellosis to develop are apparently implicated in tuberculosis infections: iodine, copper, cobalt and manganese, and, of course, the correct amounts of calcium and magnesium.

This is a notifiable disease and prevention is definitely better than cure.

Urinary Calculi

Urinary calculi (also stones in the ureter, etc.) will be prevented by seeing that the stock are on the correct minerals. If stones should arise, usually cider vinegar will help them to dissolve, as will large doses of oral vitamin C (sodium ascorbate) on occasion.

These can be a real problem in areas where the water supply is from highly mineralized bore water and bulls are especially susceptible. However, if some way can be found of including a little cider vinegar in the food, the stones do not occur. The amount needed is surprisingly small, in the case of a bull, a dessertspoon every few days would be enough.

Affected animals show signs of pain when trying to pass water and, in some cases, the pain can be so great that the beast will become very ill. Definitely a case of prevention being much easier than cure. A calcium to magnesium imbalance can also contribute to the problem.

Warts

Warts can occur anywhere on an animal, but they are more usually found around eyelids, lips, sheaths, udders — in fact, any hairless area. While not fatal, they can be a nuisance and perhaps cause debility. They are caused by a virus which apparently is only interested in a magnesium deficient host and extra dolomite in some form or other is enough to halt their progress. Once the cow is started on a daily mineral lick, the warts usually take nine or 10 days to drop off. Magnesium orotate could be used, about 4,000 mg a day (the tablets for humans usually come in 400 mg doses).

This does not work for people because they do not make their own vitamin C as cattle do (about 30 grams a day), and it seems to be needed to complete the cure. So if an animal was very poor, some vitamin C drenched in with the magnesium would help.

Woody Tongue

This is exactly what it says, the tongue goes hard and the animal cannot eat. The sovereign remedy in the old days was to get the poor wretch into market before it starved in order to save something out of the wreck. Occasionally the hardness can be seen around the back of the tongue and the side of the head as well, accompanied by swelling.

It is caused by an organism called actinomycosis which likes iodine-deficient hosts. Stock that have been having their licks do not seem to succumb and I have not seen it for many years now.

The old treatment was to fetch the vet and have sodium iodide injected straight into the vein. I have seen equally quick results from a brew of four tablespoons seaweed meal, mixed with a liter of cider vinegar shaken up well in a bottle (it is a slightly explosive mixture and the lid should be left loose), 250 ml of which is drenched into the sufferer every few hours. In the case I cured, the cow recovered by the time the bottle was finished. This was before the days of liquid seaweed which could certainly be given instead.

Worms and Liver Fluke

Drench resistance strikes fear into everyone these days, but it seems to be a fact of life. The worms adapt to drenches faster than we can make new ones. Even the Ivermectin group, which was supposed to be proof against drench resistance, has now succumbed. Each new drench has a limited life as long as it is of a chemical composition.

The answer to worms lies in good husbandry which has been outlined in earlier chapters. We shall never be able to beat the worms, so we must use an organic system of farming that lets the dung beetles, earthworms and soil fauna do it for us. This must be allied to a diet high enough in the necessary minerals to stop the worms from becoming a scourge. Dr. William A. Albrecht says in his works that animals who have the right amount of copper in their systems do not have worm problems. Farmers who have fully remineralized their land and have it in good heart have, in many cases, given up drenching on a regular basis. Most of them also see that their cattle have licks available when they want them.

Given the copper information found in Chapter 8, where it is pointed out that Dr. Albrecht found worm infestations (of any kind) only occurred in copper-deficient animals, the section below on different kinds of worms is academic. It has been noticeable with all stock that fluke, tape worm and coccidia are the first and easiest to prevent with even quite small amounts of copper. Those farmers with cattle on the lick described in Chapter 8 will find that drenching becomes a thing of the past.

A worm count of 200 or below is not a concern in properly supplemented and fed animals; in fact, a "wormless" beast usually is not very well since worms do not live in unhealthy animals. Not only actual worms, but all protozoa-type infections appear to be caused by a lack of copper in the diet. It took me and other farmers a few years to realize that many of the conditions, such as coccidia and possibly toxoplasmosis, just were not occurring once the ration was in order.

As the copper in the lick prevents the worms from staying in the gut of the cattle, they will surely die out fairly soon as they have to live inside a beast to complete their life cycle. It is interesting that worm counts done soon after beasts arrive often show a quite high count of eggs, but no adults either mature or immature. Another week or two on the lick is probably enough to see that the animal is fully clear. When hatched, worms just do not stay in an animal whose copper reserves are correct.

Copper and worms

I have not used any proprietary drenches for just on 30 years now. Copper sulfate, with various additions, was used for many years prior to the advent of artificial chemical drenches in the late 1950s. The copper was mixed with either carbon tetrachloride (a very poisonous cleaning fluid), lead arsenate (another dangerous poison) or nicotine sulphate, which was possibly the safest of the three. I very much doubt if the reported deaths were often due to copper poisoning.

Copper toxicity causes liver damage which, if not treated, is fatal. We found out that when copper is administered with dolomite, there is little risk unless the cattle have been grazing heliotrope or some other weed high in copper like Patterson's Curse or St. John's Wort; however, if they had, the chance of a worm burden would be virtually nil, because of the high copper content of all three.

According to the Department of Primary Industries in Queensland, the blood serum levels of copper in a bovine should be between 500 and 1,100 milliliters per liter, at which levels worm infestations would be unlikely. In all cases of suspected worm infestation, a count should be taken either by the vet, or as many farmers the world over these days do, examining the manure with a microscope (a school quality microscope will do).

Long-standing copper overload can apparently be corrected by giving the affected cow dolomite on a permanent basis. This can be given with an injection of vitamin B15 (Pangamic acid) (10 cc) and vitamin C (20 cc) in the same syringe once a week. This has been tried in the field on farms where too much copper has been spread on the land. For immediate copper poisoning, give the beast a tablespoon of dolomite and vitamin C powder by mouth every few hours, and 10 cc of vitamin B15 with 30 cc of sodium ascorbate (vitamin C) in the same syringe by injection. This can be given every few hours, although a calf that I first did this work on recovered fully in an hour and a half and further doses were not necessary. Signs of copper poisoning are misery and a hunched up appearance — in effect, acute belly ache due to liver pain.

According to Justine Glass, black animals need about six times as more copper than white ones. Consult the section on copper for deficiency signs.

Initially several friends who ran cattle, horses, sheep or goats experimented using copper instead of proprietary drenches, with very satisfactory results. The only controlled experiment was performed with goats and the Department of Agriculture did the tests. Half were given the latest state-of-the-art drench (not Ivermectin), and half were given the copper sulfate/dolomite/vitamin C dose. The results were equal — 100 percent clear in both cases.

When I first started using copper sulfate instead of proprietary drenches, I could not find any guidelines and Dr. Albrecht, whose works show that copper prevents worm infestations, does not mention dosage quantities. A retired vet lent me a copy of the *British Veterinary Codex* (1952) and I was able to work out amounts from that source. I had reckoned that monogastrics need about half the amount on body weight that ruminants require; however, work done by the University of Minnesota on ponies and copper requirements suggests that equines actually top the list as far as copper requirement or tolerance goes.

I have discussed running copper through the diet with various vets and at least one did not have apoplexy, but was genuinely impressed and interested because to use his exact words, "We have come to the end of the line with proprietary drenches." That was 18 years ago and the situation has not improved with the years.

Modern strategy

I found that telling people to drench their animals with copper and dolomite, etc., was not a success, but what *does* work is prevention as underlined many times in this book.

Working with various farmers we evolved the lick mentioned in Chapter 8. It has been eminently successful and it is no longer necessary to think about the old strategic drenching. It takes a full year to build up the copper reserves in an animal and only then does the coat stay a good strong color the whole year through. The cattle's continuing good health on the regime seems to be the only consequence.

Past strategy

What follows is purely academic and a relic of the drenching days. That said, there are a few things farmers should know about the types of worms that used to lie in wait for their stock.

Barber's Pole Worm (Haemonchus contortus)

Without a doubt the most insidious and dangerous worm on the list. This is a blood-sucking worm that can be picked up and carried in encapsulated form until the time is right for it to emerge in the animal's gut and wreak havoc. It is particularly deadly in calves and kills by totally robbing its host of red blood cells. It is not active when the weather is cold, but waits for a really hot spell to emerge, when it causes almost total anemia very quickly. Drenching when it is in the encapsulated form does not work, so it is necessary to drench the moment a young animal's lower eyelids become pale; by the time they are white, it is usually too late. Another case where prevention in the form of good farming practice is the best strategy.

Brown Stomach Worm (Ostertagia)

Another blood-sucking worm, but this can be hit whenever the animal is drenched. Quite often two drenches are needed on successive days, as the worm burrows into the stomach walls, and does not come out until those at large have been destroyed by the first drench. One very badly infested animal that I leased needed drenching for four days in succession before he was clear.

Lung Worm (Meullerius capillaris and Dirofilaria immitis)

These are two distinct sorts of lung worm that used to need two different drenches. It was also thought that they did not cover more than one sort of livestock, but this does not appear to be correct.

Signs are coughing and ill thrift. A worm count will be needed to determine which kind of lung worm is causing the trouble, unless using the Ivermectin group of treatment. Lung worm, if not dealt with, causes lasting damage. Quite often really healthy looking animals with a slight cough do not have lung worm, but have scarring from a previous infestation. Doses of vitamin E, as for pneumonia scarring, could possibly help.

If calves are badly infected with lung worm, using a severe drench can be fatal because all the worms in the lungs will be killed at once. This will mean there will not be enough room left for the animal to gain the oxygen it needs, so it virtually dies of mechanical pneumonia. It is better in this case to use a mild drench that deals with the worms in the intestines (part of the lung worm cycle is spent there), and then a day or two later use something that will kill those left in the lungs, some of which will have migrated by then.

Liver Fluke

Fluke has a six-week life cycle from a small conical snail whose larvae infect the stock. These can infest very wet land or dams and, once they have been ingested, grow into a full-grown (about a centimeter wide) fluke in the host's liver.

Signs of liver fluke are anemia, ill thrift, occasionally a swelling under the jaw and a capricious appetite. Liver fluke drenches are very expensive, and rather severe. Raising the copper in the ration until the eyelids become a deep pink usually gets rid of fluke. This could mean a flat dessertspoon per cow for a few days, after that make sure the copper levels are correct and the fluke will not re-infest the animals. If they do, the beasts are not getting enough copper.

The drenches for fluke are very expensive and often difficult to get. However, it is the easiest to prevent; a fairly small amount of copper run through the ration is enough to prevent it. I did not realize this when I first went onto an irrigation farm (the fluke come in with the water), but wondered why I was the only person on the system that was not losing animals with fluke. The animals were on a small maintenance ration of copper in those days. Farmers have since told me that they put a thick canvas bag of copper sulfate by the water outlet onto the farm so that small amount is washed into the water at all times. That was their way of preventing fluke. Treating dams for fluke with copper sulfate is not always the answer, because the snails can and do migrate across damp paddocks. It is not necessary to have a dam to have problems with liver fluke infestations.

Pin worms (Nematodes)

These are not often a problem in cattle, but occasionally they can infest a calf that would pick them up from licking the ground. Examining the calf around the anus will reveal them squirming. A Piperazine drench will be needed to get rid of them, the normal drenches (barring the Ivermectin group) do not usually touch them. Piperazine powder is used for chooks and the dosage can be determined from the container. I found that pea hay was a potent source of pin worm infestation, the eggs must have been scraped off the ground with the pea straw. Again the lick is the answer.

Tapeworm (Moniezia expanza)

There is some controversy on tapeworms. Some authorities say they are species specific, in other words, cattle could not pick up tapeworms from a dog, but others say this is rubbish. Tapeworms in stock are not very common, so perhaps they really do not cross between species. The white segments sometimes can be seen in manure. If tapeworms are present, often the animal is very potbellied. A special drench is needed, on which the vet will advise. Apparently they are not killed by the Ivermectin group. One sheep farmer who had been regularly dosing them with this group decided to try a copper drench. He gave each of them two grams of copper sulphate per head in a large teaspoon of dolomite. To his amazement, many of them passed quite long tapeworms. He was unaware that they had them and had thought that the drench he used would have kept them worm-free. Copper is regarded as being a very old specific against tapeworms.

Resources Cited

Acres U.S.A. Various Issues. Available from P.O. Box 91299 Austin, TX 78709-1299, 1-800-355-5313, e-mail *info@acres usa.com*, website *www.acresusa.com*.

Adams, Ruth. *Complete Home Guide to All the Vitamins*. Larchmont Books: New York, 1971.

Albrecht, Dr. William A. *The Albrecht Papers*, volumes 1-4. Acres U.S.A.: Austin, 1975.

Allenstein, Dr. "Cow-side Practice: Molds Are Causing Feeding Problems." *Hoard's Dairyman*, 1993.

"Anthelmintic Supplement." *Farmer's Weekly*, May 1997.

Auerbach, Charlotte. *Science of Genetics*, revised edition. Hutchinson Press: London, 1969.

Bairacli Levy, Juliette de. *Herbal Handbook for Farm and Stable*. Faber and Faber: London, 1952.

Balfour, E. B. *The Living Soil and the Haughley Experiment*, revised edition. Universe Books: New York, 1976.

Begley, Sharon. "The End of Antibiotics." *Newsweek*, 1994.

Bellfield, Wendell O. "Megascorbic Prophylaxis and Megascorbic Therapy: A New Orthomolecular Modality in Veterinary Medicine." *Journal of the International Academy of Preventive Medicine*, volume 2, 1975.

Blood, D., Henderson and Radiostits. *Veterinary Medicine.* Baillierre Tyndall: London & Sydney, 1979.

Caldwell, Gladys. "Fluoride." *Acres U.S.A.*, March 1989.

CSIRO (Commonwealth Scientific Industry Research Organization). *Rural Research Bulletins* [Australia], various issues.

Davis, Adelle. *Let's Get Well*, 6th edition. Allen and Unwin: United Kingdom, 1972.

Everist, Selwyn P. *Poisonous Plants in Australia.* Angus and Robertson, 1981.

Fukuoka, Masanobu. *The One Straw Revolution.* Rodale Press: PA, 1978.

Fukuoka, Masanobu. *The Road Back to Nature: Regaining Paradise Lost.* Japan Publications: Tokyo & New York, 1987.

Glass, Justine. *Earth Heals Everything: The Story of Biochemistry.* Peter Owen: London, 1958.

Goodman, Louis S. and Gillman, Alfred. *The Pharmacological Basis of Therapeutics*, 5th edition. Macmillan Publishing: New York, 1970.

Gregg, Charles T. *The Plague, an Ancient Disease in the Twentieth Century*, revised edition. University of New Mexico Press: Albuquerque, 1986.

Grieve, Mrs. M. *Modern Herbal.* Penguin Books: London, 1976.

Hensel, Julius. *Bread from Stones.* Acres U.S.A.: Austin, 1991.

Howard, Albert. *Agricultural Testament.* Oxford University Press: New York and London, 1943.

Hubbard, C.E. *Grasses*. Penguin Books: London, 1972.

Hungerford, T. *Hungerford's Diseases of Livestock*, 8th edition. McGraw-Hill Book Company: Sydney, 1951.

Jarvis, D. C. *Folk Medicine*. White Lion Publishers: London, 1958.

Jensen, Bernard and Anderson, Mark. *Empty Harvest*. Pavely Publishing Group: Garden City, 1973.

Johnson, Clarence. "Twenty Day Shelf Life in Fluid Milk." *American Dairy Review*, July, 1979.

Kalokerinos, Archie. *Every Second Child*. Thomas Nelson: Australia, 1974.

Kalokerinos, Archie, Dettman, G. and Dettman, I. *Vitamin C, Nature's Miraculous Healing Missile*. Veritas Press: Queensland, 1993.

Kalokerinos, Archie and Dettman, G. "Vitamin C and Cancer." Transcript of an address given to the International Association of Cancer Victims and Friends, Melbourne, Australia, March 31, 1979.

Kelly, Robin A. *The Sky Was Their Roof*. Andrew Melrose: London, New York & Toronto, 1955.

Kervran, Louis C. *Biological Transmutations*. Happiness Press: CA, 1980.

Kessler, J. "Elements Mineraux Chez le Chevre, Donne de Base et Apports Recommands." Paper presented at ITOVIC INRA, International Symposium on Feeding Systems for Goats, Tours, France, 1981.

Kettle, P. R., Vlassof, A., Reid, T. C. and Horton, C. T. "A Survey of Nematode Control Measures Used by Milking Goat Farmers and of Anthelmintic Resistance on Their Farms." *New Zealand Veterinary Journal*, 1981.

Kinsey, Neal and Walters, Charles. *Hands-on Agronomy*. Acres U.S.A.: Austin, 1993.

Klenner, Frederick R. "Observations on Dose and Administration of Ascorbic Acid When Employed Beyond the Range of a Vitamin in Human Pathology." *Journal of Applied Nutrition*, Winter 1971.

Lamand, M. "Metabolisme et Besoins en Oligo-elements des Chevres." Paper presented at ITOVIC INRA, International Symposium on Feeding Systems for Goats, Tours, France, 1981.

Leahy, J. *Use and Abuse of Drugs*. Publication of the University of Western Australia.

Marston, Hedley, and Robertson, Brailsford. "Utilization of Sulfur by Animals." *CSIR Bulletin*, no. 39, 1928.

McCabe, E. *Oxygen Therapies, A New Way to Approach Disease*. Energy Publications: Morrisville, NY, 1988.

Mackenzie, David. *Goat Husbandry*. Faber & Faber: London, 1957.

McKenzie, Ross A. *Studies on Calcium Deficiencies in Horses Induced by Grazing Tropical Grasses Containing Oxalates and Testing Supplementation in the Field*. Publication of the Queensland Department of Primary Industry, 1984.

McKenzie, Ross A. and Dowling, R. M. *Poisonous Plants, A Field Guide*. Information Series, no. Q192035. Department of Primary Industries: Brisbane, Australia, 1993.

MacLeod, F. *A Veterinary Materia Medica and Clinical Repertory*. C. W. Daniel and Company: Saffron, Walden: 1995.

"Moldy Corn Poisoning in Horses." *Acres U.S.A.*, February 1995.

Moore, James. *Outlines of Veterinary Homeopathy for Horse, Cow, Dog, Sheep and Hog Diseases*, 7th edition. Henry Turner and Co.: London, 1874.

Moskowveitch, Richard. "Immunizations: A Dissenting View." Lecture 8 from *Dissent in Medicine: Nine Doctors Speak Out.* Contemporary Books: Chicago, 1984.

Neilsen, Forrest H. "Boron, an Overlooked Element of Potential Nutritional Importance." *Nutrition Today*, January/February 1988.

Passwaters, Richard A. *Selenium as Food and Medicine: What You Need to Know.* Keats Publishing: New Canaan, 1980.

Pickering, Robert. Correspondence with author, 1997.

Reynolds, E. F. *The Extra Pharmacopoeia*, 28th edition. The Pharmaceutical Press: London, 1982.

Schuessler, W. A. *Biochemic Handbook.* New Era Laboratories: London.

Shepparton Veterinary Clinic Newsletter (Victoria, Australia), April 1997.

Stone, Irwin T. *The Healing Factor: Vitamin C Against Disease.* Grosset and Dunlap: New York, 1972.

Turner, Newman. *Fertility Farming.* Acres U.S.A.: Austin, 2008.

Udzal, F. A. and Kelly, W. R. "Enterotoxemia, Pulpy Kidney Disease of Goats." Paper presented at Seminar 96, Banyo Conference Centre, Brisbane, Queensland.

Voisin, André. *Grass Productivity.* Island Press: Washington, D.C., 1959.

Voisin, André. *Soil, Grass and Cancer.* Acres U.S.A.: Austin, 1999.

Volker and Steinberg. "The vitamin requirements of goats." Paper presented at the ITOVEC INRA Symposium International, Tours, France, 1981.

Wallach, Joel and Lan, M. *Let's Play Doctor!* Double Happiness Publishing Co. 1997.

Walters, Charles. *Eco-Farm — An Acres U.S.A. Primer*. Acres U.S.A.: Austin, 1992.

Walters, Charles. *Weeds: Control Without Poisons*. Acres U.S.A.: Austin, 1996.

Whitby, Coralie. "How Do Soils Affect our Diet?" Talk presented to the Orthomolecular Association of Australia Seminar, June 25, 1982.

Widdowson R. Lecture at Kiewa Valley Seminars, 1982.

Willis, Harold. "Roots." *Acres U.S.A.*, December 1994.

Yeomans, P. A. *The Challenge of Landscape: The Development and Practice of Keyline*. Keyline Publishing Company: Sydney, 1958.

Yiamouyiannis, John. *Fluoride, the Aging Factor*. Health Action Press: Delaware, OH, 1993.

Index

Droughtmaster, 5, 14, 20, 91
drugs, to avoid, 129-136
dry feed, intake, 69-70
dystocia, 44, 152; and potassium
    deficiency, 106

Eco-Farm, 49
eczema, 152; and zinc, 109
embryo transplants, 132
Empty Harvest, 58
energy value, of feed, 69
enterotoxemia, 130, 153
environment, and breeding, 57-59
enzyme systems, and magnesium,
    95
erythromycin, 178
eucalyptus, 173
Everist, Selwyn, 173
Every Second Child, 136
eye health, and vitamin A, 112-113

facial eczema, 181
farm, efficient set up, 27-36; stress
    areas, 30-31
farming, chemical, 9; organic, 8-10
fat, as feed, 70; in meat and milk,
    4-5
feed, dusty, 73; moldy, 74; pellets,
    71; requirements, 65-77
feedlots, 7
fencing, 3-4, 27-28
fertility, and copper, 99; and
    molybdenum, 104; and seleni-
    um, 107; and vitamin A, 112;
    and vitamin E, 122
Fertility Farming, 162
fertilizer, high-nitrogen, 46
flag, 153
flax, 176
fluke, liver, 197-198
fluoroacetate, 182
fodder, management of, 34-35
foot and mouth disease, 154
foot rot, 154-155; and copper, 99
foot scald, 155
founder, and magnesium
    deficiency, 96
fracture, compound, 159-160
Friesian, 14-15, 63; British, 13

fungal conditions, and copper, 99
fungi, 74
fusarium, 74

garlic, 128-129
gates, 27-28
Gelbvieh, 21, 25
Goat Husbandry, 161
goat's milk, for calves, 87
goitrogenic feed, 70; and iodine
    deficiency, 102; and selenium,
    108
Goode, John, 101
grass, African-type, 76
grass seed abscess, 138-139
grass tetany, 155-156
grazing, cell, 3; intensive, 3
Guernsey, 13, 15
gypsum, 35, 41, 52

Hands-On Agronomy, 38
hardware disease, 156-157
Haughley Experimental Farm, 8-9
haulm, 176
hay, 34-35
Hayes, Jervis, 32
healing, and arnica, 127
heliotrope, 173-174; and copper, 48
Helms, Douglas, 29
*Herbal Handbook for Farm and
    Stable,* 125
heredity, 57-59
Hereford, 5, 13, 20, 21
Herriot, James, 183
Highland, 20-21
Hill, Stuart, ix
Hoard's Dairyman, 74
hormones, 131-134; and vitamin A,
    113
house milkers, 80-81; handling of,
    81; ration for, 82-83
Hungerford's Diseases of Livestock,
    100, 144, 151
hydrogen, 48-49
hydrogen peroxide, and mastitis,
    168

idophor, 33
Illawarra Shorthorns, 5, 13, 25

remedies, 137-199
remineralizing, cost of, 55-56
reproduction, and BGH, 133
reproductive system, and zinc, 109
retinol, 112-114
rhubarb, 175
riboflavin, 116
rice pollards, as feed, 70
rickets, 120-121, 185
ringworm, 185; and cider vinegar, 126; and copper, 99
rock dust, 2-3
Ross River fever, 186
rye grass, 177

Salers, 22
salinated land, 30
Salvation Jane, 174
sandy blight, 170-171
Sanga, 21, 23
scabby mouth, 186
Scientific American, 148
scouring, 99; and copper deficiency, 71
scrapie, 146
seaweed, 72-73, 111, 141; for calves, 87; and copper, 101; as iodine source, 102; and natural sodium, 108; and selenium, 107; and zinc, 109
selenium, 43, 106-108; and breeders, 62; deficiency, 107; and infertility, 157-158; and vitamin E, 122
Selenium as Food and Medicine, 107
Science of Genetics, The 106
shock, and arnica, 127
Shorthorn, 13, 14, 20, 21, 25
Simmental, 25
Sims, Fletcher, 36
skin, and sulfur, 109
slurry, use, 35-36
snakebite, 169, 186-188; and vitamin C, 119-120
sodium, 44-45, 108; intake, 67
sodium ascorbate, 119-120
sodium fluoride, effects on minerals, 95

soft teeth, and magnesium deficiency, 96
soil analysis, 37-56; defined, 38-39
soil organisms, 47
solids-not-fat, 4-5
sorghum, 177
Spanish windlass, 85
spider bites, 188
spirodesmin, 74, 181
St. John's Wort, and copper, 47-48, 175
staggers, and vitamin B1, 115
steroids, 131-132
stilbestrol, 132
sudex, 177
sulfur, 9, 71, 72, 108-109, 190; and breeders, 62; and garlic, 128; and parasites, 167; and ticks, 18
sunburn, and vitamin H, 123
superphosphate, 47, 54, 75; and calcium depletion, 94; and cancer, 149; and cattle health, 68; and copper deficiency, 100-101; and grass tetany, 156; and magnesium inhibition, 95; and phosphorus, 105
Suttle, Neville, 93
swelling, and BGH, 134; and comfrey, 128
SWEP, 38

tapeworm, 198-199
tendon damage, 161
tetanus, 158, 188-189; and vitamin C, 119
tetany, lactation, 165; and magnesium deficiency, 96
thiaminase, 115, 181
thiamine, 115
three day sickness, 190
thyroid, and iodine, 101-102
tick bite, 189-190
tick fever, 184
ticks, 18; and sulfur, 109
toxemia, 65; in pregnancy, 182-183
toxoplasmosis, 190

Metric Conversions

Linear Measurement

Meters x 3.281 = Feet

Meters x 39.37 = Inches

Centimeters / 2.540 = Inches

Kilometers / 1.609 = Miles

Weights & Volume

Liters / 3.785 = Gallons

Liters / 0.9463 = Quarts

Liters / 0.02957 = Ounces

Grams / 28.349 = Ounces

Kilograms / 0.4536 = Pounds

Liters x 1000 = Cu. Centimeter

Area

1 Sq. Kilometer x 247.105 = Acres

1 Sq. Meter x 1.19599 = Sq. Yards

Other Books by Pat Coleby . . .

Natural Goat Care

BY PAT COLEBY

Goats thrive on fully organic natural care. As natural browsers, they have higher mineral requirements than other domestic animals, so diet is a critical element to maintaining optimal livestock health. In *Natural Goat Care,* consultant Pat Coleby shows how to solve health problems both with natural herbs and medicines and the ultimate cure, bringing the soil into healthy balance. Topics include: correct housing and farming methods; choosing the right livestock; diagnosing health problems; nutritional requirements and feeding practices; vitamins and herbal, homeopathic and natural remedies; psychological needs of goats; breeds and breeding techniques. *Softcover, 374 pages. ISBN 978-0-911311-66-2*

Natural Horse Care

BY PAT COLEBY

Proper horse care begins with good nutrition practices. Chances are, if a horse needs medical attention, the causes can be traced to poor feeding practices, nutrient-deficient feed, bad farming and, ultimately, imbalanced, demineralized soil. Pat Coleby shares decades of experience working with a variety of horses. She explains how conventional farming and husbandry practices compromise livestock health, resulting in problems that standard veterinary techniques can't properly address. Natural Horse Care addresses a broad spectrum of comprehensive health care, detailing dozens of horse ailments, discussing their origins, and offering proven, natural treatments. *Softcover, 184 pages. ISBN 978-0-911311-65-5*

Natural Sheep Care

BY PAT COLEBY

A comprehensive guide for all breeders of sheep, whether for wool, meat or milk. Coleby draws on decades of experience in natural animal husbandry to provide essential information for both organic and conventional farmers. The original edition has been expanded significantly in the areas of breeding for finer wool and meat, land management, sheep management and treatment of health problems. Coleby covers breeds of sheep, wool, meat and milk production, feeding requirements, poisonous plants, land management, minerals and vitamins, herbal, homeopathic and natural remedies, and more. *Softcover, 232 pages. ISBN 978-0-911311-90-7*

Treating Dairy Cows Naturally

BY HUBERT J. KARREMAN, V.M.D.

In this groundbreaking work Dr. Karreman invites us to journey in to the world of dairy cows from a truly holistic perspective. With much thought and research, he builds a foundation from which to view dairy cows as animals that occupy a unique agro-ecological niche in our world. From within that niche, he describes how cows can be treated for a wide variety of problems with plant-derived and biological medicines. Drawing upon veterinary treatments from the days before synthetic pharmaceuticals, and tempering them with modern knowledge and clinical experience, Dr. Karreman bridges the world of natural treatments with life in the barn in a rational and easy to understand way. In describing treatments for common dairy cow diseases, he covers practical aspects of biologics, botanical medicines, homeopathic remedies, acupuncture and conventional medicine. By incorporating conservation principles, he also alerts us to the need of keeping our waterways clean — both for our health and the health of the cows. This book should serve as a useful reference for years to come. *Hardcover, 412 pages. ISBN 978-1-60173-000-8*

The Keys to Herd Health

BY JERRY BRUNETTI

Whether dairy or beef, a healthy herd begins in such keystone concepts as biodiversity on the farm, acid/alkali balance in feedstuffs, forage quality, and more. In this accessible video, eco-consultant and livestock feed specialist Jerry Brunetti details the keynote essential for a successful livestock operation. A popular speaker at eco-farming events across North America, Brunetti explains the laws of nature in terms farmers can embrace, and doles out specific steps you can utilize on your farm right away — all in a convenient video format that you can watch and review whenever you like. *DVD & VHS, 57 min. ISBN 978-0-911311-91-2*

Cancer, Nutrition & Healing (Video)

BY JERRY BRUNETTI

In this remarkable video presentation, Jerry shares the priceless lessons and wisdom gained through his successful struggle with an aggressive form of lymphoma. You'll never look at cancer — or cancer treatments — in quite the same way after viewing Jerry Brunetti's step-by-step plan for restoring health using fresh foods and natural, holistic, herbal treatments. You will learn about: strengthening immunity; holistic treatment protocols; health-boosting recipes; supplements and detoxification; supplemental conventional therapies; foods to eat; foods to avoid; and much more! *DVD & VHS, 85 min. ISBN 0-911311-81-5 (DVD) / ISBN 0-911311-82-3 (VHS).*

The Non-Toxic Farming Handbook

BY PHILIP A. WHEELER, PH.D. & RONALD B. WARD

In this readable, easy-to-understand handbook the authors success-fully integrate the diverse techniques and technologies of classical organic farming, Albrecht-style soil fertility balancing, Reams-method soil and plant testing and analysis, and other alternative technologies applicable to commercial-scale agriculture. By understanding all of the available non-toxic tools and when they are effective, you will be able to react to your specific situation and growing conditions. Covers fertility inputs, in-the-field testing, foliar feeding, and more. The result of a lifetime of eco-consulting. *Softcover, 236 pages. ISBN 978-0-911311-56-3*

Weeds: Control Without Poisons

BY CHARLES WALTERS

For a thorough understanding of the conditions that produce certain weeds, you simply can't find a better source than this one — certainly not one as entertaining, as full of anecdotes and homespun common sense. It contains a lifetime of collected wisdom that teaches us how to understand and thereby control the growth of countless weed species, as well as why there is an absolute necessity for a more holistic, eco-centered perspective in agriculture today. Contains specifics on a hundred weeds, why they grow, what soil conditions spur them on or stop them, what they say about your soil, and how to control them without the obscene presence of poisons, all cross-referenced by scientific and various common names, and a new pictorial glossary. *Softcover, 352 pages. ISBN 978-0-911311-58-7*

Eco-Farm: An Acres U.S.A. Primer

BY CHARLES WALTERS

In this book, eco-agriculture is explained — from the tiniest molecular building blocks to managing the soil — in terminology that not only makes the subject easy to learn, but vibrantly alive. Sections on NP&K, cation exchange capacity, composting, Brix, soil life, and more! *Eco-Farm* truly delivers a complete education in soils, crops, and weed and insect control. This should be the first book read by everyone beginning in eco-agriculture . . . and the most shop-worn book on the shelf of the most experienced. *Softcover, 476 pages. ISBN 978-0-911311-74-7*

To order call 1-800-355-5313
or order online at *www.acresusa.com*

Acres U.S.A. — books are just the beginning!

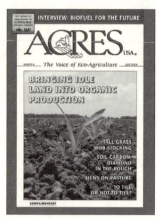